# *Regaining to Know Aright*
## *'Natural' Knowledge for a Secular World*

### Geoffrey Wilkinson

Copyright © Geoffrey Wilkinson 2024

Geoffrey Wilkinson has asserted his right under the Copyright, Designs and Patents Act 1988 to be identified as the author of this work. All rights reserved. No part of this publication may be reproduced, stored in a retrieval system, or transmitted in any form or by any means, electronic, mechanical, photocopy, recording or otherwise, without prior written permission of the copyright owner. Nor may it be circulated in any form of binding or cover other than that in which it is published and without a similar condition including this condition being imposed on a subsequent purchaser.

A CIP catalogue record for this book is available from the British Library.

ISBN 978-1-9160622-1-4

Published by Geoffrey M. Wilkinson
Franksbridge
Llandrindod Wells
Powys LD1 5SA
United Kingdom

for Susan and Bryn

# Contents

*To the reader*

| | |
|---|---:|
| 1. Introduction: scope and method of the essay | 3 |
| 2. What we have lost, when we lost it, and a name for what we must recover in its place | 5 |
| 3. The fortuitous concourse of all things: Lucretius celebrates the world of Epicurus and the Atomists | 20 |
| 4. No reasonable definition of reality: Einstein, Bohr, and the 'incompleteness' of quantum mechanics | 28 |
| 5. Thrown into the world: early Heidegger on 'Being-there', 'the nothing', and 'resoluteness' | 37 |
| 6. Nothingness and suchness: the Sixth Patriarch, *śūnyatā*, and Nishida's notion of 'the good' | 63 |
| 7. Conclusions: time we knew the world aright | 83 |
| Sources | 106 |
| Notes | 113 |

## *To the reader*

I suppose the nagging suspicion of having missed a turning somewhere, or gone about things the wrong way, is common enough in individual human experience. Many people must have discovered for themselves that there are situations in which a seemingly absurd logic — 'If I were going where you're going, I wouldn't start from here' — makes perfect sense. What, though, if the logic sometimes applies not just within the span of a single life but to all of us, collectively and historically? Do whole societies, cultures, traditions miss turnings or go about things the wrong way? The premise of this essay is that they do, and that the Judaeo-Christian tradition has done so by insisting, falsely, that the world exists as it does because it expresses the will and wisdom of a creator God. If that sounds intriguing as a premise, or provocative, or otherwise of interest, I hope you will read on. If not, not.

The idea for the essay grew out of a journal article, 'The Frog and the Basilisk', first published in 2015. Some of the material is adapted or updated from an earlier book, *Certainty, that thing of indefinite approximation*. Full publication details of both titles can be found in my Sources.

# 1. Introduction: scope and method of the essay

For as long as anyone can document or deduce, and who knows for how much of the larger history of *Homo sapiens* as a species, we have tended to believe that the world we inhabit is intelligible. Intelligible, that is, in the sense of having some reason for being as it is, or for existing at all — which is not to say that we have always expected to understand the reason, only that we have assumed there must be one. Since the early twentieth century, however, cosmology and physics have confronted us with robust and growing evidence that in fact our world is *un*intelligible. For it is made of the stuff of the universe, energy and matter that burst into being spontaneously, discontinuously, without reason, necessity or purpose at an arbitrary instant 13.8 billion years ago, and stuff that at the subatomic level confounds our very conception of reality. There is a profound mismatch here between old habits of mind and new evidence, and it is high time we resolved it.

The title of the essay alludes to John Milton's tract *Of Education*, which declares that the end of learning is 'to repair the ruins of our first parents by regaining to know God aright'. For Milton and for Francis Bacon, both devout Protestants, repairing the ruins demanded that we undo the Fall by studying God's works in the world and recovering the 'natural' knowledge that man had possessed before he ate the forbidden fruit. But for secular learning, and a secular society, the 'natural' knowledge we need to recover is not of the world as evidence of a creator God and his purposes. It is of the world itself, uncreated, purposeless, and ultimately unintelligible.

At first sight this seems as daunting a project as undoing the Fall. Yet far from being discouraged, we would do well to remember that there have been philosophers and others who did not assume intelligibility to be a fact of nature and who, each in their own way, offer us the prospect of a more honest accommodation with the world. The essay considers four of them in turn:

- The Roman poet Titus Lucretius Carus (died c. 55 BC), who transformed Atomist physics into an exuberant vision of life lived most fully when we recognize that the world has come into being at random, not by divine fiat or design.

- The Danish physicist Niels Bohr (died 1962), whose reply to the charge that quantum mechanics gives an 'incomplete' description of physical reality was, in effect, that the concept of physical reality is meaningless at the subatomic scale.
- The German philosopher Martin Heidegger (died 1976), who observed that our most characteristic relation with the world is not that of a subject among objects but of finding oneself 'thrown' there, 'Being-already-amidst-the-world'.
- The Japanese philosopher Nishida Kitarō (died 1945), whose lifetime endeavour to know and express 'true reality' was a moral or, as he would say, 'religious' project as much as a philosophical one.

Lucretius (primary source his great didactic poem, *De rerum natura*) is overtly prescriptive, both urging us to see the world anew and telling us how we ought to live accordingly. Bohr (various sources, relating especially to his exchanges with Einstein) is very exact and exacting about the language to be used in describing quantum phenomena but when he speaks about the educational implications, for example, his tone is one of patient advocacy rather than prescription. Heidegger (*Being and Time* and *What is Metaphysics?*) and Nishida (*An Inquiry into the Good* and other works) are essentially non-prescriptive in that they want to bring about a radical transformation in our understanding of the world and yet do not spell out what difference the transformation could or should make to the way we live our lives. Similarly, this essay asks what is the *character* or *quality* of a knowledge that is genuinely 'natural' and has the potential to transform our understanding of the world, without attempting to prescribe how it should change the way we live. That is partly because 'natural' knowledge does not necessarily make any difference to the day-to-day practicalities of our lives. More fundamentally, it is because any difference 'natural' knowledge does make, practical or otherwise, is a matter of individual conscience and individual choice, for which there is no prescription.

## 2. What we have lost, when we lost it, and a name for what we must recover in its place

> After all, everything we believe, even the most far-reaching cosmological theories, has to be based ultimately on common sense, and on what is plainly undeniable.
>
> Thomas Nagel, *Mind and Cosmos: Why the Materialist Neo-Darwinian Conception of Nature Is Almost Certainly False* [1]

On the face of it, the assumption that everything we believe has to be based ultimately on common sense seems, well, common sense. But on reflection, and as I hope to demonstrate, it is just that, an assumption, not a self-evident truth, and it is strange to find it endorsed so matter-of-factly by the respected moral philosopher Thomas Nagel. As the subtitle of his book forewarns, Nagel regards the 'reductive materialism' of the neo-Darwinian orthodoxy as 'a heroic triumph of ideological theory over common sense.' [2] 'It is prima facie highly implausible,' he writes, 'that life as we know it is the result of a sequence of physical accidents together with the mechanism of natural selection.' [3] He is adamant that Darwinian evolution alone cannot account for the appearance in the world of consciousness, reason, and especially value, and this is the starting point for his exploration of 'alternatives that make mind, meaning, and value as fundamental as matter and space-time in an account of what there is.' [4] An atheist himself, Nagel discounts divine intervention as an alternative, which leads him to the conclusion — tentative yet for him evidently more in accord with common sense — that an 'expanded, but still unified, form of explanation will be needed', one that will include Darwinian evolution and the laws of physics and chemistry, but may have to include 'teleological elements' as well. [5] So, we have a contemporary philosopher convinced that common sense is an indispensable *and reliable* guide to understanding the cosmos and our place in it. Really? What is going on here, and how does it relate to 'the larger project,' as Nagel calls it, 'of making sense of the world'? [6]

Let's begin at the beginning, with a creation myth. Every culture has one and I could probably start with almost any of them. I choose, however, to start with a hybrid myth: the Judaeo-Christian fused with aspects of Plato and the neo-Platonist philosopher Plotinus.

The Book of Genesis tells an incomplete story: while it describes in splendid, day-by-day detail *how* God created the world, it does not explain *why* he did it as he did, or did it at all. It says, for example, that God put man in the garden of Eden to dress and keep it, but what was his motive for making the garden in the first place? On the basis of Genesis alone, without the benefit of later theological and philosophical glosses, we are left wondering whether God may have made the world on a whim, because he could. The creation myth to be found in Plato's *Timaeus*, on the other hand, both explains how the creator framed our world (modelling it on the world of the eternal Forms)[7] and gives a motive, a good motive, for his doing so. It is an inferior world, Plato insists, one of change and becoming, a poor imitation of the 'real' world the creator had before him as he worked. Nevertheless he created our world because, as Plato has Timaeus phrase it, 'He was good, and what is good has no particle of envy in it; being therefore without envy he wished all things to be as like himself as possible.'[8] We now have a motive for the creation, then, but also another question: how exactly does an absence of envy produce a world? Plotinus, writing in the third century AD, answered this question by turning Plato's benevolent creator into a generative principle called 'the One' (or 'the Good'), which is 'perfect because it seeks for nothing, and possesses nothing, and has need of nothing; and being perfect, it overflows, and thus its superabundance produces an Other.'[9] Albeit produced indirectly, not directly, by the 'superabundance' of the One, the world is

> a sort of Life stretched out to an immense span, in which each of the parts has its own place in the series, all of them different and yet the whole continuous, and that which precedes never wholly absorbed in that which comes after.[10]

Fortuitously, Plotinus's image of the world as an 'immense span' (which was elaborated by the theologians and philosophers of later centuries into a great scale or 'chain' of being)[11] offers a way into the larger notion I am trying to get at: the supposed intelligibility of the world.

For something to be intelligible it has to be 'reasonable' or, as Nagel would say, 'make sense'. How does that affect our perceptions of the world? Presumably (and beyond the scope of this essay) for reasons to do with the way our cognitive abilities have evolved, it seems that the world must satisfy certain expectations, or demands, before it can make sense to us. One of them is that it should be *orderly*. Now Plotinus's image expresses the complexity and fullness of the world, but it also clearly implies that there is a place for everything, everything should be in its place ('its own place in the series'), and yet all still mutually consistent and coherent ('the whole continuous'). Given the sheer range of Plotinus's influence on theology and philosophy — from early Christian, Judaic and Islamic doctrine through to the writings of nineteenth-century German idealists — I think he provides a particularly significant illustration of our demand for orderliness and consistency, together with closely related qualities such as continuity, regularity, and predictability. Moreover, it is a demand that applies to every feature of the world and how it came into being, and in an extreme case it may have surprising consequences. One extreme case is that of the philosopher Leibniz, who was so concerned about orderliness and continuity that he was prepared to deny the theological doctrine of God's absolute free will. It is a denial that results from the interaction of two of Leibniz's principles, 'sufficient reason' and 'perfection'. In plain language, sufficient reason asserts that there must be some reason why everything is as it is (or exists at all) rather than otherwise (or not at all). Leibniz locates every such reason in an order of eternal truths, a 'metaphysical' necessity incapable of further explanation. The upshot is that this metaphysical (or 'logical') necessity ultimately determines the existence, attributes and behaviour of everything in the world, even if it appears to be a world that might never have come into being or that might have come into being differently. In other words, an order of metaphysically necessary reality determines our familiar and merely contingent order of reality.

'Besides the world or aggregate of finite things,' Leibniz wrote in 1697, 'we find a certain Unity which is dominant.... [in the sense that it] not only rules the world, but also constructs or makes it; and it is higher than the world and, if I may so put it, extramundane; it is thus the ultimate reason of things. For neither in any one single thing, nor in the whole aggregate and series of things, can there be found the sufficient reason of existence.'[12] He goes on to argue that the aggregate of finite things is derived from eternal truths by means of 'a certain Divine mathematics or metaphysical mechanics'.[13] It soon emerges that the dominant Unity (or 'Being of metaphysical necessity') is God, who must exist to provide not just the ultimate reason of things but also the mechanism by which eternal truths can be actualized in the world as 'truths that are temporal, contingent, or physical'.[14] God operates as this mechanism by choosing from among an infinite number of possible worlds ('the infinite combinations of possibles') and creating the best possible world, which, in accordance with Leibniz's quantitative and qualitative definition of perfection, is the one 'by whose means the greatest possible amount of essence or possibility is brought into existence.'[15] But notice the constraint that Leibniz has just imposed on his God: God is free to consider all possible worlds, yet he is not free to choose or create a world other than the best. He cannot go for the second best, or the third best, or the sixty-fifth best. The constraint is embodied in Leibniz's principle of perfection (the principle Voltaire parodied in *Candide* as Doctor Pangloss's doctrine that 'all is for the best in the best of all possible worlds'), which seems to exalt Leibniz's definition of the best possible world into an objective standard independent of the will of God. God cannot be permitted to define the bestness of the best possible world spontaneously because there is no guarantee that even he, God, may not have recourse to chance. Leibniz is quite blunt about this:

> If the will of God did not have for a rule the principle of the best, it would either tend towards evil, which would be worst of all; or else it would be in some fashion

> indifferent to good and evil and guided by chance. But a will which always allowed itself to act by chance would scarcely be of more value for the government of the universe than a fortuitous concourse of atoms, with no God at all. And even if God should abandon himself to chance only in some cases and some respects, ... he would be imperfect, as would the object of his choice; he would not deserve to be wholly trusted; he would act without reason in those cases, and the government of the universe would be like certain games, half a matter of chance, half of reason.[16]

God is at the inexplicable apex of sufficient reason, the principle that says there must always be a reason why anything is as it is rather than otherwise. God himself, however, must always have a reason why he wills as he does rather than otherwise and that reason is determined by a principle, Leibniz's perfection, which is not of God's own making. For it is inconceivable that the government of the universe should be in the hands or the mind of a God who played games of chance with it.[17] Such a God *would not deserve to be wholly trusted*. It is hard to avoid the conclusion that Leibniz allows God only an ambiguous, conditional sort of free will because the alternative would be to confront the intolerable notion that in fact we inhabit a world of blind chance.[18]

While the detail of Leibniz's metaphysics and theodicy may now be of interest mainly to historians of philosophy, Leibniz himself marks one of the high points in our trust, or faith, in the intelligibility of the world. In the early eighteenth century, that is, the world still made sense because theology and most of the philosophy and natural science of the day reassured us that it was divinely ordered and, in principle at least, susceptible to human understanding.

If that was a high point, where do we look for the low point?

## When the world ceased to be intelligible, and why

Leibniz's demand for order and consistency at every level and in every constituent of the universe has a further consequence: a preoccupation with gaps and leaps, or rather, with proscribing gaps and leaps under a 'law of continuity'. 'Nothing takes place all at once,' Leibniz pronounces,

> and it is one of my most important and best verified maxims that *nature makes no leaps*. This ... *law of continuity* ... means that the passage from the small to the great and back again always takes place through that which is intermediate....[19]

Nature makes no leaps, indeed. Leibniz is in for a shock. And so are we. The reality is that nature *does* make leaps, literally. The passage from the small to the great and back again does not always take place through that which is intermediate. Sometimes it takes place all at once, randomly and discontinuously. I mean the so-called quantum jump, a phenomenon of nature at the microscopic or subatomic scale, as opposed to the macroscopic or classical scale. How the two scales are defined and how they are separated (or are they separated?) are among questions taken up later in the essay and I will say no more about them for now. But the quantum jump itself tells a story that deserves to be better known, and is powerfully metaphorical.

In 1986, independently and within months of each other, two teams of physicists in the United States and a third in Germany successfully demonstrated the existence of quantum jumps. More precisely, they observed a single atom (trapped in an electromagnetic field within a vacuum chamber) switching discontinuously between three discrete energy levels. The researchers achieved this by monitoring the random on-off blinking of a laser-excited fluorescence, measurable by a photon-counting device and, in one of the experimental setups, visible to the eye through an improvised microscope.[20] A steady fluorescence was the effect of extremely rapid (up to about one billion times per second) and indistinguishable jumps up and down between the ground state of the atom and one of the two higher-energy excited states. The on-off blinking was the effect of the relatively long (more than 30 seconds in the case of a barium ion) and detectable delay before the atom spontaneously returned to its ground state from the second of the excited levels; until it did so, the atom was not 'available' for the rapid kind of jumping and therefore the fluorescence was quenched. Both effects derived from the emission or absorption of a photon, a packet or quantum of light, that occurred with each downward or upward jump. The transitions from one energy level to another did not take place through that which is intermediate as they involved electrons, which can jump from one well-defined energy state to another but nowhere 'in between'. There is no intermediate state in the atom because, as Einstein correctly proposed in 1905 as an explanation of the photoelectric effect, electromagnetic energy is quantized, that is, has to be thought of in terms of discrete quanta that have no 'in-between' amounts.[21] Although the quantum jump is just one phenomenon of nature at the subatomic scale (which from now on I refer to as the quantum scale), metaphorically, I suggest, it could represent all physical phenomena that confound our conception of what is reasonable or makes sense.

Unintelligibility at the quantum scale is mirrored — and writ very large — at the cosmological scale. Now, of course, I mean the unintelligibility of the Big Bang, the primordial origin of the universe. The phrase 'Big Bang' is unfortunate because it conjures up and distracts us with images of a cosmic fireworks display, whereas the true significance of the event itself lies not in the impressiveness of the Bang but in the fact that it happened at all. The universe has not existed for ever: it came into being at a particular and arbitrary point in time (more strictly, *together with* time).

The evidence trail begins in 1915 with Einstein's general theory of relativity and the problem he subsequently had reconciling his own equations, which indicated that the universe must be either contracting or expanding, with his preconception of a static universe.[22] Einstein's modification of the equations with the so-called cosmological constant compensated for the gravitational attraction of matter that would otherwise cause a finite universe to collapse and had the desired effect, as he thought, of preserving a static universe. In 1931 Einstein removed the cosmological constant from his equations, on the grounds that it was redundant or, as he may or may not have said, 'my biggest blunder'.[23] But in the meantime, and one of the stranger ironies in the history of science, a Catholic priest, Georges Lemaître, had grasped what Einstein himself had not: in a French-language research paper published in Brussels in

1927, little noticed at the time, Lemaître observed that Einstein's constant belonged in the equations and there was no reason to remove it, and yet it did not preserve a static universe.[24] On the contrary, as we now understand it, the cosmological constant produces not just an expansion but an *accelerating* expansion of the universe.[25]

The first observational evidence of the expansion of the universe — and the reason Einstein abandoned his cosmological constant — dates to 1929, when Edwin Hubble demonstrated that light coming from distant galaxies is red-shifted (shifted towards the long-wavelength end of the spectrum), indicating that they are rapidly moving away or receding. Some four years earlier, and employing Henrietta Leavitt's technique for measuring the distance to a type of star known a Cepheid variable,[26] Hubble had definitively proved that the nebulae, which were thought to be clouds of gas within our own Milky Way galaxy, are in fact separate galaxies outside and far beyond the Milky Way. That was shock enough, because it was generally believed that the Milky Way comprised the whole of the universe. But the even greater shock that Hubble delivered in 1929, building this time on the work of fellow-astronomer Vesto Slipher and assisted by Milton Humason, was that the red-shift of a receding galaxy is directly proportional to the galaxy's distance from earth.[27] In other words, the further away a galaxy is, the faster it is receding. To cut a long story short, not only is it a well-established cosmological fact that the universe is expanding, we now have a pretty good idea (and certainly a much better idea than Hubble himself had) of how fast it is expanding: this value, known as the Hubble constant, puts the recession velocity at about 70 kilometres per second per megaparsec, which means that a galaxy 1 megaparsec (about 3.26 million light-years, the typical separation distance between galaxies) from the earth is moving away from us at approximately 70 kilometres per second.[28]

What of the Big Bang itself? What is the evidence for it, and how does it fit into the current picture, the standard cosmological model (based on general relativity *with* the cosmological constant as it is now understood), of the origin and evolution of the universe? The most obvious, almost ludicrously obvious evidence for the Big Bang requires no more than a simple analogy and simple logic. Things that are violently flying apart from each other in all directions — the fragments of shrapnel from an exploding bomb, say — must, at any given moment in the past, have been closer together: run the sequence backwards and they get closer and closer together until eventually they are all more or less in the same place. In the case of an exploding bomb, it would only take a few seconds or fractions of seconds to wind the sequence back to the juncture at which everything was in the same place. In the case of a universe exploding out of nothing, it would take a lot longer to wind back the sequence, but the principle is the same. If the universe is expanding, and it is, at some juncture in the past everything that constitutes it must have been closer together. And not just closer together, not just more or less in the same place or even exactly the same place, but compressed into a single point that is infinitesimally small, and unimaginably dense and hot.

Less intuitive as evidence of the Big Bang, yet validated empirically, is the distribution of the lightest elements, hydrogen, deuterium, helium and lithium, throughout the observable universe. From about one second after the Big Bang, it is calculated, the temperature would have fallen to 10 billion kelvins (10 billion degrees on the Kelvin scale), allowing protons and neutrons to bind together to form the nuclei of the light elements. This phase is known as the Big Bang nucleosynthesis. Because such nuclear reactions are well understood from laboratory experiments here on earth, it is possible to predict statistically what nuclei would have been produced, and in what proportions, in the nucleosynthesis. The surprise prediction is that while, for example, roughly 25 per cent (by mass) of the primeval protons and neutrons would have gone to form helium, very few went into the nuclei of atoms heavier than lithium: the heavier elements were synthesized in nuclear processes within stars, and the first stars did not appear for about another 100 million years. The surprise prediction is impressively borne out by observation, which makes it, according to the theoretical physicist Lawrence Krauss, 'one of the most famous, significant, and successful predictions telling us the Big Bang really happened. *Only a hot Big Bang can produce the observed abundance of light elements and maintain consistency with the current observed expansion of the universe.'*[29]

The third major piece of evidence in direct support of the Big Bang is the cosmic microwave background radiation diffused throughout the universe. Theory predicts that the last heat of the Big Bang, by now cooled to about 3 kelvins (3 degrees above absolute zero), should be coming at us from all directions in the sky in the form of microwave radiation. That is precisely (albeit accidentally) the signal detected in 1964 by the radio astronomers Arno Penzias and Robert Wilson: sure enough, what Penzias and Wilson had detected, and for which they were later awarded a Nobel Prize, was the predicted afterglow of the Big Bang. Subsequent observations of the microwave background radiation, including several using instruments mounted on high-altitude balloons, have since been repeated with vast improvements in accuracy and certainty by two orbiting satellites: an American satellite known as WMAP, launched in 2001, and the Planck satellite launched by the European Space Agency in 2009. All these observations have succeeded in identifying small variations in the temperature and polarization of the background radiation ('hot spots' and 'cold spots', or anisotropies) at different points in the sky. It is hard to overstate the significance of this achievement, because the patterns of the variations make it possible to reconstruct a picture of the universe as it would have been some 380,000 years into its existence — not quite all the way back to the Big Bang, but long before the formation of stars and galaxies.[30]

One measure of how rapidly and clearly the evidence of the Big Bang has come into focus is that little more than three decades ago the total age of the universe could only be estimated somewhere in the range 10-20 billion years.[31] At the time of writing, thanks to the Planck satellite, the age of the universe is confidently dated at 13.8 billion years (more correctly, $13.797 \pm 0.023$ billion years).[32] While broadly consistent with the age ($13.772 \pm 0.059$ billion years) calculated in 2012 using WMAP data, the margin of uncertainty of 23 million years compares favourably with the WMAP uncertainty of 59 million years.[33] Moreover, confidence in the Planck estimate was further bolstered in 2020 when it was found to be in generally good agreement with independent observations made by the ground-based Atacama Cosmology Telescope in Chile.[34]

Early in the eighteenth century, Leibniz assured us that God had been guided by perfection when he set about the creation; hence, by definition, the fabric of the world he had made, and the fabric of the universe itself, was perfect, unblemished by chance-happening, arbitrariness, or anything else that might affront the human sense of reasonableness. Today, early in the twenty-first century, it is difficult to decide which is the greater affront to reasonableness — electrons that flit randomly between discrete energy states (and other quantum objects that do other, even weirder things, as we shall see), or a whole universe that spontaneously bursts into existence out of nothing, discontinuously, without reason, necessity, or purpose. But then I've already answered that conundrum: they are two scales of one and the same affront, namely that the government of the universe *is* left to a fortuitous concourse of atoms, with no God at all.

The chronology of how we have got to our current state of discomfiture stretches back more than one hundred years, and can be plotted by some of the key dates along the way, momentous dates, though not always recognized as momentous at the time:

- **1905** Einstein explains the photoelectric effect in terms of the quantization of energy, clarifying an earlier and poorly understood suggestion by Max Planck, and laying a foundation stone of quantum physics;
- **1917** Einstein modifies his theory of general relativity with a constant that he thinks will preserve his preconceived picture of a static universe, only to discover a few years later that in fact the constant produces an expanding universe (and, had he lived another 50 years, would have discovered that it is expanding at an accelerating rate);
- **1927** In a French-language journal mostly overlooked in the English-speaking world, Lemaître points out Einstein's mistake in removing the constant from general relativity and confronts cosmology with the reality that, if the universe is expanding, reversing the expansion takes it back to some

juncture — a *beginning* — when its entire contents were concentrated in one infinitesimally small point;
- **1929** Hubble formally publishes his observational evidence of the expansion of the universe, findings on the red-shift of galaxies, demonstrating not only that the galaxies are receding from us and from each other, but that the further away they are the faster they are receding;
- **1964** Penzias and Wilson stumble upon microwave background radiation at about 3 degrees above absolute zero across the sky — the faint afterglow of the Big Bang and a critical piece of observational evidence that it happened as theorists model it;
- **1986** Three teams of experimental physicists independently observe the fabled quantum jump, and on each occasion Leibniz spins inconsolably in his grave;
- **2020** Results published by the ground-based Atacama Cosmology Telescope in Chile agree with, and are generally taken as independent confirmation of, the observations of the Planck satellite, which has made the most accurate measurements to date of the cosmic microwave background radiation and narrowed down the age of the universe to 13.8 billion years.

I doubt that Professor Nagel would dispute the factual basis of the chronology. But where, Professor, is the common sense in it?

## A name for what we must recover: 'natural' knowledge

The title of this essay alludes to the following passage in Milton's tract *Of Education*, first published in 1644:

> The end ... of learning is, to repair the ruins of our first parents by regaining to know God aright, and out of that knowledge to love him, to imitate him, to be like him, as we may the nearest, by possessing our souls of true virtue, which, being united to the heavenly grace of faith, makes up the highest perfection. But because our understanding cannot in this body found itself but on sensible things, nor

arrive so clearly to the knowledge of God and things invisible as by orderly conning over the visible and inferior creature, the same method is necessarily to be followed in all discreet teaching.[35]

I am drawn particularly to Milton's phrase 'regaining to know' (rather than 'gaining to know') because it conveys the idea that we have been here before — not, I hasten to add, in the biblical sense that we once enjoyed a state of innocence which we then lost through an act of disobedience, but in the simpler, non-doctrinal sense that it is never too late to relearn something we seem to have forgotten. That we have been here before and relearning are recurring themes of the essay.

Written in the turbulent years of the English civil war, the strenuous educational programme set out in *Of Education* has practical and political aims, including to ensure that in future 'political societies ... may not ... be such poor shaken uncertain reeds, of such a tottering conscience as many of our great councillors have lately shown themselves.'[36] Yet the radical aim of Milton's programme is, as he stresses, to 'repair the ruins of our first parents', which means nothing less than *undoing the Fall* by restoring the state of knowledge that man had enjoyed before he ate the forbidden apple. That is to say, restoring the state in which Adam had named the beasts of the field when God brought them to him — 'and whatsoever Adam called every living creature, that *was* the name thereof,' as Genesis puts it.[37] According to Martin Luther, Adam was able to do so because of his 'superior knowledge and wisdom', that is, 'solely because of the excellence of his nature, [by which] he views all the animals and thus arrives at such a knowledge of their nature that he can give each one a suitable name that harmonizes with its nature.'[38] Luther characterizes the quality of Adam's knowledge in terms of completeness and perfection: it is 'superior' knowledge. By subtle contrast, Francis Bacon characterizes its quality in terms of 'naturalness'. What enabled Adam to choose his names for the beasts, says Bacon, was 'the pure light of natural knowledge', 'natural' in the sense that it was unadulterated by any temptation to acquire 'moral' knowledge, the knowledge appropriate only to God:

> it was not that pure light of natural knowledge, whereby man in paradise was able to give unto every living creature a name according to his propriety, which gave occasion to the fall; but it was an aspiring desire to attain to that part of moral knowledge which defineth of good and evil, whereby to dispute God's commandments and not to depend upon the revelation of his will, which was the original temptation.[39]

Bacon's message is that we profoundly misunderstand the Fall if we think it means that our proper state is one of abject ignorance, or that the reality of God's works must remain mysterious and unexamined. In our fallen condition we 'dispute God's commandments', but that is the fault of the wilful uses to which we put our knowledge, not of the knowledge itself. There is nothing wrong with knowledge. Bacon is turning theology into natural philosophy, or the beginnings of empirical science, as we would call it (which is another story well beyond my scope). At the same time, however, he is urging us to study God's works in the world around us, and to study them as evidence of his wisdom and benevolence, just as Milton commends 'orderly conning over the visible and inferior creature' as the only way to arrive 'clearly' at a knowledge of God. The knowledge we have to regain, according to Bacon and Milton, is 'natural' and 'clear', but not disinterested: it is knowledge of God and his purposes, not knowledge for its own sake.

It may seem perverse to invoke Bacon and Milton when my whole intention is to argue that the end of secular learning is to know the world in and of itself, for its own sake, not as evidence of an imagined creator. And yet I am also drawn to that phrase '"natural" knowledge', in the sense of knowledge of or relating to nature (or the world as a whole), because, like 'regaining to know', it suggests that what we need is there waiting to be recovered — with the additional connotation that it *is* a disinterested knowledge. The task I am proposing is akin to undoing the Fall, except that what is to be recovered in the undoing is not 'natural' knowledge in a theological sense but knowledge of the world itself, in its own right, *and whether intelligible or not.*

It looks an impossible task. Historically and culturally, individually and collectively, we are conditioned to regard the world as something we can and must make sense of, and we struggle to assimilate the irrefutable, scientific evidence that a lot in the world does not make sense. Hapless victims of the empirical outlook that Bacon did so much to encourage, we find ourselves caught in a pincer movement of our own discoveries at either end of the scale: the very big (cosmology) and the very small (quantum physics). But what is the problem? Is it that the world is unintelligible, or that we go on expecting it to be intelligible? After all, who or what says that the world has to make sense to us? It is disingenuous at best to ignore the findings of well founded and well documented scientific enquiry, or to pigeonhole them as 'difficult' and of no consequence to us. On the contrary, I believe that the mismatch between our entrenched assumptions about the world and the factual evidence now available is, or should be, of consequence to us. We must unsee the world as we think we know it and try to see it anew. Or rather, and it is the method of this essay, try to see the world through the eyes of four people who have demonstrated in different ways that it is possible to find an accommodation with the unintelligible.

## 3. The fortuitous concourse of all things: Lucretius celebrates the world of Epicurus and the Atomists

The Latin-speaking poet Lucretius lived in the first century BC. He is known for a single work, a great didactic poem in six books called *De rerum natura*, which translates literally as *On the Nature of Things* or, probably closer to the nuance Lucretius intended, *On the Nature of the Universe*. It seems a good place to start:

> I will now set out in order the stages by which the initial concentration of matter laid the foundation of earth and sky, of the ocean depths and the orbits of sun and moon. Certainly the atoms did not post themselves purposefully in due order by an act of intelligence, nor did they stipulate what movements each should perform. But multitudinous atoms, swept along in multitudinous courses through infinite time by mutual clashes and their own weight, have come together in every possible way and tested everything that could be formed by their combinations. [40]

For those of us brought up on the biblical creation story, Lucretius's version of events comes as quite a surprise, or shock. 'In the beginning,' Genesis tells us, 'God created the heaven and the earth.' St John concurs: 'All things were made by him; and without him was not any thing made that was made.' Even other pagan philosophers agreed that a divine agency was involved in the making of the world. Think, for example, of the creator at work in Plato's *Timaeus*. Or of Aristotle's God, who, seemingly too aloof in his own perfection to have created the world, nonetheless acts as the Unmoved Mover that sets the world in motion. Not so, Lucretius insists, the universe came into being of its own accord, without cause, purpose, design, or designer: in the beginning was 'nothing but a hurricane raging, a newly congregated mass of atoms of every sort.'[41] What we see and experience as our world, he says, is the result of those atoms coming together 'in every possible way', for over endless time they have rushed about, colliding and combining, sometimes breaking apart again, testing anything and everything that could ever be formed — the sun, the earth, sky, ocean, and eventually ourselves and all the other living creatures. Lucretius does not deny the existence of gods, but he does deny that they had anything to do with the creation of the world, and he heaps particular scorn on the notion ('sheer nonsense') that they arranged the world for man's convenience: are we to suppose, he scoffs, that nature could not manage the changing seasons, which enable man to grow crops and benefit in other ways, without divine guidance? [42]

For Lucretius, all this talk of colliding and combining is literal truth, not metaphor, because he subscribes to the Atomist theory of physics first propounded by Leucippus and Democritus in the fifth century BC, and developed by another Greek philosopher, Epicurus, in the fourth century BC.[43] Like our everyday word 'atom', the term Atomism relates to the Greek adjective '*atomos*', which translates as 'uncuttable' or 'indivisible'. According to the Atomist theory itself, everything that exists comprises just two fundamental constituents: an infinite number of indivisible atoms, which are invisible and constantly in motion, and an infinite volume of empty space. While the atoms themselves are indestructible and eternal, and while the gods are the one exception to the rule,[44] the myriad varieties of compound matter that the atoms create are perishable. It is not only the human body (and soul) that must die; in time, say Epicurus and Lucretius, even the earth and the stars will come to an end, and the atoms that formed them will be scattered all over again to career about in the infinite void.

Now the intention of Epicurus and Lucretius is not, as we might think, to drive us to despair by demonstrating that we inhabit an arbitrary, precarious and meaningless universe, and that therefore our lives are arbitrary, precarious and meaningless. On the contrary, Epicurus and Lucretius are out to demonstrate that the world is full of wonder and delight, and that we have every reason to value and enjoy our lives to the full. But to understand how and why they try to do so, first we need to make a detour.

One of the recurring themes of *On the Nature of the Universe* is that everything has come into existence *randomly*, not as the result of any deliberate or inevitable process. In arguing, for instance, that infinite atoms in infinite space must have produced innumerable (presumably infinite) other worlds besides our own, Lucretius repeats that 'our world has been made by nature through the spontaneous and casual collision and the multifarious, accidental, random and purposeless congregation and coalescence of atoms'.[45] He stresses the quality of randomness because he needs to counter the rigid determinism inherent in the earliest Atomist physics and philosophy. Here he follows Epicurus, who, deploring the 'inescapable and merciless necessity [i.e., determinism]'[46] of Democritus, had come up with an ingenious means of escaping it — the '*clinamen*' or 'swerve'. Epicurus agrees that the atoms are in constant motion but, where Democritus has them all move straight downwards under their own weight, he claims that at unpredictable moments some of them veer off slightly, 'swerve', from their downward trajectory. One reason Epicurus introduces this device is to address a logical inconsistency: if, as he points out, atoms of different weights fall uniformly at the same speed (an assumption he has no way of substantiating, yet which we now know to be true of macroscopic objects falling in a vacuum), there can be no opportunity for them to collide and coalesce into compound matter, and thus no way for nature to create anything. His main reason, however, and it is made even more explicit in Lucretius's poem, is to account for the phenomenon of free will in living creatures. The Atomism of Democritus implies that the behaviour of animate matter is as mechanistic as the behaviour of inanimate matter, for there can be no scope for acts of choice or will if everything is determined by the motion of atoms and their 'everlasting sequence of cause and effect'.[47] Neither Epicurus nor Lucretius explain exactly how the 'swerve' is supposed to enable free will. Nevertheless, the device is crucial to their larger intention, to which we now return.

Free will is so important to Epicurus and Lucretius because they have a moral purpose, one that in effect amounts to a programme of moral re-education. Which is to say, they intend their philosophy to change how we think and behave — impossible without free will. For Epicurus, philosophy is a serious business that has to be judged by its practical, almost therapeutic, results: a philosopher who fails to bring solace to the mind, he is said to have protested, is as useless as a physician who fails to cure a disease of the body.[48] Lucretius evidently shares Epicurus's earnestness. He, too, would have us feel differently about the world and our place in it. Poet as well as philosopher, though, he chooses to disguise his purpose in his art: perhaps adapting Epicurus's imagery, he likens himself to a physician who tricks a child into drinking a foul-tasting medicine by smearing the rim of the cup with honey.[49] Yes, Lucretius is saying, I offer a nasty medicine that will shake your most cherished assumptions, confront you with your greatest fears. But it is for your own good, and in the end you will be cured of your delusions and your fears, free to live a better life than the one you live now. How much better, we ask, or whose life should we aspire to? Lucretius's answer — another surprise or shock to us — is *the life of the gods*. Just as the gods did not create our world, so they have no interest in what we get up to in the world, and it is precisely because they are untroubled by human affairs that they enjoy an existence of 'utter tranquillity.... free from all pain and peril'.[50] The gods, in short, are a paradigm of the serenely contented life that we can and should enjoy. 'True piety,' according to Lucretius, 'lies ... in the power to contemplate the universe with a quiet mind.'[51] Elsewhere he writes, 'there is nothing to prevent men leading a life worthy of the gods',[52] which may be an allusion to Epicurus's 'you will live as a god among men.'[53] Epicurus should know, after all: although a mortal human being, his philosophy has made a god of him.[54]

The overarching moral principles of Epicureanism, then, are peace of mind and the pursuit of happiness. As Epicurus pithily expresses it, 'one must practise the things which produce happiness, since if that is present we have everything and if it is absent we do everything in order to have it.'[55] By 'the things which produce happiness' he does not mean — as his contemporary and later detractors were quick to allege — sensuous and unbridled hedonism, but rather 'the lack of pain in the body and disturbance in the soul'.[56] The sources of happiness, it turns out, are understated, even austere: barley cakes and water, not the flesh and wine of an extravagant feast; a simple life, not one perturbed by appetites that demand ever more elaborate satisfactions,[57] or driven by greed and the pursuit of personal fame, wealth and power. The watchwords of Epicureanism are not indulgence and excess. They are prudence and modesty. Prudence, Epicurus says, is 'the greatest good', which is why it is 'a more valuable thing than philosophy. For prudence is the source of all the other virtues'.[58] Likewise Lucretius: 'if a man would guide his life by true philosophy, he will find ample riches in a modest livelihood enjoyed with a tranquil mind.'[59] To have a lot will always be to crave more; to have a little will be to be content with little, and of that there can be no lack.

It seems odd now that *On the Nature of the Universe*, which embodies such an essentially well-meaning philosophy, should ever have offended anybody — odder still given that the text was effectively lost for more than a thousand years after Lucretius's death. Yet it did give offence from the start, and the offence blew up all over again, so to speak, after the text was rediscovered in a German library in 1417.[60] Even in his own troubled times, when Rome descended into factional rivalry and intermittent tyranny, many of Lucretius's sentiments must have caused outrage. 'Far better to lead a quiet life in subjection than to long for sovereign authority and lordship over kingdoms',[61] for example, was unlikely to appeal to a proud people with an ethos of glorious conquest. Most telling for our purpose is the offence, for which the evidence persists at least into the seventeenth century, that Lucretius caused to Christianity. *On the Nature of the Universe*, all the more suspect for the seductive quality of its poetry, posed a brazen challenge to the doctrines, the institutions and, above all, the moral authority of the church. There were attempts to suppress the poem altogether. In 1516, for instance, study of it was banned in Florentine schools, and in 1549 it came close to being included in the list of books (known from slightly later as the *Index Librorum Prohibitorum*) forbidden by the church authorities. But the church's tone and preferred method of self-defence had already been established long before, in the fourth century AD, when St Jerome recorded that Lucretius was driven mad by a love potion and committed suicide at the age of forty-four.[62] By insinuating that *On the Nature of the Universe* was the work of madman and, for good measure, that parts of it had to be corrected by Cicero, Jerome sought to create — or, more likely and more bluntly, fabricate — an image of Lucretius calculated to discredit him.

That Lucretius was pagan was not a problem in itself. So were Plato and Aristotle, yet Christian doctrine had found ways of accommodating them. The problem was that, unlike Plato and Aristotle, Lucretius denied any divine agency in the creation of the universe. According to the physics he adopted from the Atomists and Epicurus, the gods — or God — could have had no hand in even the most fundamental constituents of the universe, the atoms, which had a completely independent and eternal existence of their own. Then there was Lucretius's assertion that the soul dies with the body and there is no afterlife, which threatened to undermine both church dogma and one of the cornerstones of the established moral order, namely the fear of divine judgement and endless torment after death for sinners. (The cynical might say that the denial of an afterlife was also bad for church business, because it made a nonsense of the later practice of selling indulgences, the 'credits' purported to ease the suffering of souls in Purgatory.) While Lucretius's views on the pagan religion of his own day — mere ignorance and superstition, he mocked — presented no direct threat to Christianity, the central message of Epicureanism was subversive of any organized religion, pagan or otherwise. Early in the fourth century AD, Lactantius, latterly tutor to the son of the Emperor Constantine, acknowledged that Epicureanism had a large following because 'the attractive name of pleasure invites many.'[63] And that was its offence, for to make the pursuit of happiness the greatest good, as did Epicurus and Lucretius after him, was to appeal to our all too human but all too misguided inclinations. Worse still, Lactantius argued, Epicureanism tailors its appeal to our individual weaknesses and vices. It has something for everyone: the idle, for example, are told that they have no need to exert themselves, the timid no need to do their military service, and the covetous no need to share their wealth with those less fortunate than themselves. Although he does not say so explicitly here, surely Lactantius expects us to agree that 'Please yourself' is a false and wicked injunction, and that the only true and virtuous injunction is the church's: 'Please God, not yourself', which means subordinating our own desires to God's purposes, and obeying God's commands in fear of his righteous anger.

There is a charge against Epicureanism that is especially problematic for us, and one that Lucretius's poem itself invites but does not wholly succeed in rebutting. The charge, that is, of encouraging egoism and a callous indifference to the misfortunes of others, as Lucretius appears to do in the opening lines of Book 2. The greatest joy of all, he writes, is

> to possess a quiet sanctuary, stoutly fortified by the teaching of the wise, and to gaze down from that elevation on others wandering aimlessly in search of a way of life, pitting their wits one against another, disputing for precedence, struggling night and day with unstinted effort to scale the pinnacles of wealth and power.[64]

Despite (or because of) Lucretius's unconvincing disclaimer, 'Not that anyone's afflictions are in themselves a source of delight', this passage strikes the modern reader as deeply repugnant. The heart sinks further when Lucretius invokes a sort of evolutionary history to suggest that early man was not naturally given to altruistic behaviour: the first human beings, he says, 'lived out their lives in the fashion of wild beasts roaming at large.' He goes on:

> They could have had no thought of the common good, no notion of the mutual restraint of morals and laws. The individual, taught only to live and fend for himself, carried off on his own account such prey as fortune brought him.[65]

As we begin to fear the worst, Lucretius introduces a 'social contract' theory of moral development into his evolutionary history. True, it implies, altruism was not innate in us. But we *learned* to be altruistic in the interests of self-preservation, starting from the moment when 'neighbours began to form mutual alliances, wishing neither to do nor to suffer violence among themselves.'[66] And it was from such modest, pragmatic origins that we gradually developed our notions of the common good, law, and morality. Although far from a uniform process ('It was not possible to achieve perfect unity of purpose'), the very fact that we are here to tell the tale is evidence that it worked. If it had not,

> the entire human race would have been wiped out there and then instead of being propagated, generation after generation, down to the present day.

There remains the suspicion that many, if not most, adherents of Epicureanism reserved their altruism (for which their word 'friendship' may have been a synonym) for other Epicureans, not for the mass of unenlightened humanity. In spite of this lingering doubt, however,

Lucretius's poem as a whole convinces us that, whatever its likely failings in practice, Epicureanism is not a philosophy which sanctions irresponsible egoism. Quite the reverse. It teaches that the greatest good, the pursuit of happiness, is inseparable from the pursuit of a virtuous life. There cannot be one without the other. It is succinctly put by the Greek poet and philosopher Philodemus, more or less a contemporary of Lucretius and, like him, an Epicurean. It is impossible to live pleasurably, writes Philodemus,

> without living prudently and honourably and justly, and also without living courageously and temperately and magnanimously, and without making friends, and without being philanthropic.[67]

Once again, the virtues that Epicureanism extolls do not seem in the least extraordinary or controversial to us. We have to recall, therefore, that well into the Christian era what was considered scandalous about Epicureanism was its denial that such virtues — and moral behaviour in general — derive from or require divine authority. According to Epicureans, virtue and morality begin, not end, with the recognition that there is no divine authority for them to derive from. What they do derive from is our own capacity, personally and collectively, to choose one way of behaving over another; to agree one way is desirable because, for example, it strikes the best possible balance between individual and common good (or however else we define 'right' in secular law, ethics and custom), while another is undesirable because it fails to strike that balance (or however else we define 'wrong'). In a sense we come back to the swerve, Epicurus's device to account for free will, otherwise precluded altogether by the determinism of the earliest Atomists and constrained by any recourse to divine authority. Virtue and morality come from our free will, come from us. And yet, as Epicurus would be the first to remind us, we ourselves are part of the world in which and about which we make our choices, and are made of the same material as the world. Atoms. Multitudinous atoms.

## 4. No reasonable definition of reality: Einstein, Bohr, and the 'incompleteness' of quantum mechanics

> Physics is an attempt conceptually to grasp reality as it is thought independently of its being observed. In this sense one speaks of 'physical reality'.
>
> Albert Einstein, 'Autobiographical Notes' [68]

> [A]ny observation of atomic phenomena will involve an interaction with the agency of observation.... Accordingly, an independent reality in the ordinary physical sense can neither be ascribed to the phenomena nor to the agencies of observation.
>
> Niels Bohr, 'The Quantum Postulate and the Recent Development of Atomic Theory' [69]

There must be many ways of characterizing the prolonged — and, some would say, still unresolved — debate between Albert Einstein and Niels Bohr over the philosophical significance of quantum mechanics. One way is to see it as a clash between ontology and epistemology, between the branch of philosophy that asks about the nature and forms of *what exists*, as distinct from the branch that asks about the nature, grounds and limitations of our *knowledge of* what exists. From this perspective, and at the risk of grossly oversimplifying, Einstein stands for an ontological conception of physics (and of scientific endeavour in general), while Bohr stands for an epistemological conception.

Despite his own crucial role in the early development of quantum mechanics, Einstein never abandoned his later conviction that it was deficient as a body of theory because it failed to provide a 'complete' description of the physical world or, as he put it, 'physical reality'. In other words, his belief was that there is such a thing as physical reality, susceptible of complete description, and his complaint was that the so-called Copenhagen interpretation of quantum mechanics, associated particularly with Bohr, had given up on completeness and therefore on the pursuit of physical reality. For Bohr, by contrast, the whole point was that there had been a fundamental change in what could be expected of physics at the quantum (microscopic) level, as opposed to physics at the classical (macroscopic) level governed by Newtonian laws. What had changed was that it proved impossible to apply familiar, causal descriptions to objects or events at the scale of Planck's quantum of action (the mathematical value $h$, now known more simply as Planck's constant). Writing in 1949, Bohr expressed it as follows:

> From the very beginning the main point under debate has been the attitude to take to the departure from customary principles of natural philosophy characteristic of the novel development of physics which was initiated in the first year of [the twentieth] century by Planck's discovery of the universal quantum of action. This discovery ... has indeed taught us that the classical theories of physics are idealizations which can be unambiguously applied only in the limit where all actions involved are large compared with the quantum. The question at issue has been whether the renunciation of a causal mode of description of atomic processes ... should be regarded as a temporary departure from ideals to be ultimately revived, or whether we are faced with an irrevocable step towards obtaining the proper harmony between analysis and synthesis of physical phenomena.[70]

Einstein clearly subscribed to the view that the renunciation of a 'causal mode of description' should be seen as a temporary departure from the classical ideal: find what is missing and we will have a theory that does completely describe the nature of the world. Equally clearly, Bohr subscribed to the second view, that the renunciation represented an 'irrevocable step': at the quantum level, all we can know of the world is what we manage to observe, measure, and communicate to each other unambiguously.[71]

Whether temporary or irrevocable, the need to renounce causal description — and with it causality itself, the underlying assumption that every effect has a cause, and that a precise knowledge of the present enables us to predict the future — arises because of the indeterminacy of objects and events at the quantum scale. One manifestation of their indeterminacy, critical to the story I am about to tell, is that an observer can never know, *even in principle*, both the precise position of a subatomic particle and its precise momentum simultaneously. This is the disconcerting message of Werner Heisenberg's uncertainty relations, which say that there is a small (of the order of Planck's constant, $h$) yet irreducible uncertainty in the relation between the particle's position and its momentum. The experimental consequence is that the more accurately the position is measured, the less accurately its momentum is known, and vice versa. The uncertainty relations are not an indictment of experimental shortcomings, as if the right setup would be capable of precisely measuring position and momentum at the same time. It is not that we need better apparatus or experiments, but that it is meaningless to describe a subatomic particle as having simultaneously both a precise position and a precise momentum. Or as Heisenberg himself phrased in 1927 in the paper in which he first formulated the uncertainty relations, 'Even in principle we cannot know the present in all detail.'[72]

The uncertainty relations were, and remain, glaringly at odds with classical Newtonian physics, which does allow causal description because it deals with precise and knowable information about objects and events, causes and effects, present and future, in the macroscopic world. We can predict with certainty where billiard balls, for instance, or rockets will be after a specified time, for they obey classical laws that permit precise measurements of position and momentum to be carried out simultaneously. There is, so to speak, an unbridgeable gulf between such exact classical descriptions and the sort of descriptions quantum mechanics offers, which deal only with statistical likelihood — the probability or probabilities that an electron, say, will behave in one way rather than another as it passes through a piece of apparatus.[73]

The gulf between quantum and classical physics is nowhere starker, I think, than in the story of the Einstein-Podolsky-Rosen (EPR) thought-experiment, a story that also gets to the heart of the philosophical debate between Einstein and Bohr. In 1935 Einstein published a paper jointly with two Princeton colleagues, Boris Podolsky and Nathan Rosen. As signalled by the rhetorical question of its title, 'Can Quantum-Mechanical Description of Physical Reality Be

Considered Complete?',[74] the intention of the EPR paper was to demonstrate that the description *cannot* be considered complete. The question is easily answered, the paper claims, 'as soon as we are able to decide what are the elements of the physical reality.... We shall be satisfied with the following criterion, which we regard as reasonable. *If, without in any way disturbing a system, we can predict with certainty ... the value of a physical quantity, then there exists an element of physical reality corresponding to this physical quantity.*'[75] The crux of the EPR argument (which, as of 1935, stands or falls solely by the interpretation of the thought-experiment) is that it is possible to imagine a system such that an observer could, without disturbing the system, predict with certainty the value of a physical quantity and thus satisfy the 'reasonable' criterion of 'physical reality'.

Einstein, Podolsky and Rosen envisage two particles that interact with each other and become entangled, as we would say now, before flying apart to form two systems. Even within the constraints of quantum mechanics, we will be able to measure precisely the total momentum of the two particles and the relative distance between them when they interact. Hence if, after the two particles have flown apart, we choose to measure the momentum of the first particle, we will know, precisely and without disturbing the second system, the momentum of the second particle because the total must be unchanged. In accordance with the EPR criterion, therefore, the momentum of the second particle will be shown to correspond to 'an element of physical reality'. Alternatively (not simultaneously), we could measure the position of the first particle and precisely deduce the position of the second, without disturbing the second system and thus demonstrating that the position of the second particle also corresponds to an element of reality. To put it another way, we will have done something that quantum mechanics — which insists that momentum and position cannot both be elements of reality simultaneously — is unable to explain. Which shows, in turn, that something must be missing from the quantum-mechanical description. It is incomplete.

The authors make no attempt to deny that, even in their own thought-experiment, it is impossible to predict or measure position and momentum simultaneously. On the contrary, they mean to exploit the uncertainty relations as a trap to expose an apparent absurdity in the Copenhagen interpretation. Here is how they set the trap:

> One could object to this conclusion on the grounds that our criterion of reality is not sufficiently restrictive. Indeed, one would not arrive at our conclusion if one insisted that two or more physical quantities can be regarded as simultaneous elements of reality *only when they can be simultaneously measured or predicted*. On this point of view, since either one or the other, but not both simultaneously, of the quantities $P$ [momentum] and $Q$ [position] can be predicted, they are not simultaneously real. This makes the reality of $P$ and $Q$ depend upon the process of measurement carried out on the first system, which does not disturb the second system in any way. No reasonable definition of reality could be expected to permit this.[76]

According to the Copenhagen interpretation, in other words, for the second particle to have the reality of a precisely defined momentum, or the reality of a precisely defined position, it would somehow have to 'know' which measurement — momentum or position — we had chosen to carry out on the first particle. That was unconscionable for Einstein because it would imply that the two systems influenced each other through 'action at a distance'. There is an assumption, which goes by the name of locality, that there can be no influence between spatially separated systems; in its most restrictive form, Einstein locality, it is underpinned by Einstein's special theory of relativity, which states that no signal of any kind can travel faster than the speed of light. Nevertheless, the Copenhagen interpretation implies that the two EPR systems would influence each other instantaneously and no matter how far apart they might be, whether nanometres, kilometres, or light-years. It is hard to disagree with the judgement that no reasonable definition of reality would permit such a thing.

Einstein and his co-authors had set out to demonstrate conclusively, as they thought, that a quantum-mechanical description of the EPR experiment would be incomplete. In his reply, however,

Bohr countered that only a quantum-mechanical description — in short, the Copenhagen interpretation — was capable of recognizing that their experiment involved two *complementary* properties, momentum and position, and that *the choice of which property of the first particle to measure would, according to the conditions of the experiment, determine the 'reality' of that property for the second particle.*[77]

Complementarity, a core principle of the Copenhagen interpretation, is also invoked to account for the fact that a photon or other subatomic particle may exhibit wavelike behaviour. This seeming paradox, or 'wave-particle duality', is now generally accepted to mean that 'wave' and 'particle' have to be regarded as two facets, not paradoxical but complementary, of the same entity. Complementarity is inseparable from another seeming paradox, namely that the act of observing any system involving quantum objects *changes the system*. We cannot observe from a detached, 'outside' vantage point: our experimental equipment and, through our choices, we ourselves are part of the system, observer-participants. Recalling the lecture in September 1927 in which he had first publicly elaborated on complementarity, Bohr wrote of 'the impossibility of any sharp separation between the behaviour of atomic objects and the interaction with the measuring instruments which serve to define the conditions under which the phenomena appear.'[78] That is, complementary properties appear under mutually exclusive experimental conditions, which we arrange. If we choose an arrangement designed to measure a particle property, we cannot expect to measure a wave property in the same experiment, and vice versa. Or, as in the case of the EPR thought-experiment, if we choose to measure the position of a quantum object, we cannot expect to measure its momentum in the same experiment, and vice versa.

The most striking feature of the EPR story is the sequel which neither Einstein nor Bohr lived to witness: the experiments in and since the 1970s that have tested the thought-experiment in the laboratory.[79] They include one, conducted in 1982 by Alain Aspect and his colleagues at the University of Paris-South, which measured the polarization of a pair of photons as the experimental equivalent of measuring the position or momentum of a pair of particles. The most important technical improvement in this experiment, the third by the Aspect team, allowed the experimental arrangement to be changed while the photons were in flight, and changed so rapidly that no influence could have reached one part of the apparatus from another unless it travelled faster than light — the first test to challenge locality in its most restrictive form, Einstein locality. It is of still greater import, therefore, that the third experiment confirmed the results of the earlier two, demonstrating a degree of correlation between spatially separated systems that exceeded the limit permitted by the EPR assumption of locality but agreed well with the predictions of quantum mechanics.[80] What is more, virtually every other experiment before and after 1982, including one in China in 2017 that successfully transmitted a pair of entangled photons between an orbiting satellite and two ground stations 1,203 kilometres apart,[81] has likewise been interpreted as endorsing quantum mechanics at the expense of the EPR reasoning. In some sense, the reality of the second system *does* depend upon the measurements carried out on the first system, regardless of the distance between them. While there is no suggestion that the two systems are influencing each other by means of a faster-than-light signal (which would constitute a violation of Einstein locality), they still appear to influence one another instantaneously. Or is it muddled and misleading to talk of 'influencing' at all here because, in their entangled state, the two systems remain parts of an indivisible larger whole even after they have flown apart? However we try to express it, the almost routinely reproducible results of these experiments plainly contradict the EPR 'reasonable definition' based on locality, and confront us instead with *non*-locality as a demonstrable fact of physical reality.

In recent years there has been a renewal of interest in the EPR thought-experiment, and especially in Bohr's reply to it. There is general agreement that Bohr's wording is very opaque, which is why there is little or no consensus as to what he is actually saying, or not saying.[82] One view is that he seeks, and fails, to escape the incompleteness charge by retreating into logical positivism (more specifically, verificationism, the positivist doctrine that the significance of a statement is only revealed by the method of its verification),[83] to which one counter-argument is that Bohr's defence of quantum mechanics does not depend on 'suspect philosophical doctrines' but is 'dictated by the dual requirements that any description of experimental data must be *classical* and *objective*.'[84] An alternative to these views (and largely in response to them) is the suggestion of an 'ambivalence' in Bohr's pursuit of two different aims: showing that '"complementarity tackles the situation successfully in its own terms"' and/or that '"the EPR argument is flawed in *its* own terms".' Consequently, according to the alternative critique, the reader is not sure whether Bohr's reply 'is intended to be a purely physical one ... or whether it does rely on ... a particular philosophical position. Even if the latter is the case, again it is not made clear whether it might come from complementarity or positivism or some other set of beliefs.'[85]

It is a satisfying irony that a branch of physics so concerned with indeterminacy and uncertainty should itself be the subject of so much indeterminacy and uncertainty. Nonetheless, there is a risk that the difficulties of Bohr's writing style may obscure the significance of his reply to the EPR paper, which, don't forget, he formulated four decades before the technology existed to test the thought-experiment in a laboratory. The true significance of Bohr's reply is, I think — and I stress that this is a personal interpretation of Bohr's epistemology — that he had already seen and, implicitly if not explicitly, accepted that reasonableness in all things is a demand of the human mind but not necessarily of the matter and energy that make up the rest of the universe.[86] One sentence in particular is clear and unambiguous: '[T]he *finite interaction between object and measuring agencies* conditioned by the very existence of the quantum of action [i.e., the impossibility of making a sharp distinction between the behaviour of a quantum object and its interaction with the measuring apparatus] entails ... the necessity of a final renunciation of the classical ideal of causality and a radical revision of our attitude towards the problem of physical reality.'[87] Not just a renunciation, says Bohr, a *final* renunciation. At the quantum scale, it is not meaningful to go on expecting a 'complete' description of the world in the sense that Einstein meant. Besides, what obligation does the world have to humour our need for completeness, or reasonableness, or anything else at any scale?

Einstein belongs within the tradition in natural science and philosophy that has sought to uphold, and continues to uphold, the ultimate intelligibility of everything that exists.[88] Einstein himself described this tradition as follows: 'Certain it is that a conviction, akin to religious feeling, of the rationality or intelligibility of the world lies behind all scientific work of a higher order.'[89] Although he expressed his doubts about quantum mechanics in different terms at different times,[90] surely what troubled him fundamentally was that it posed a direct challenge to that very conviction. And yet, on the basis of our current knowledge, we have to say that quantum mechanics has confounded Einstein: at the quantum level, the intelligibility of the world can no longer be an article of absolute faith. In the specific case of the EPR thought-experiment, which is where my story began, it is difficult to see how Einstein could regard the laboratory demonstrations of non-locality as consistent with any 'reasonable' definition of reality. Faced with the compelling evidence of Aspect and everyone else who has successfully done the experiments, would Einstein finally recant his faith in 'physical reality'? Or would he shake his head and protest that the quantum-mechanical description of the world is still incomplete, refusing even from beyond the grave to forgo the search for a complete description?

# 5. Thrown into the world: early Heidegger on 'Being-there', 'the nothing', and 'resoluteness'

> Man alone of all beings ... experiences the marvel of all marvels: that what is *is*.
>
> Martin Heidegger, *What is Metaphysics?* postscript (1943) [91]

> In the end an essential distinction prevails between comprehending the whole of beings in themselves and finding oneself in the midst of beings as a whole. The former is impossible in principle. The latter happens all the time in our experience.
>
> *What is Metaphysics?* main text (1929) [92]

I confess that I approach Heidegger with some disquiet. It is partly because I have only the most rudimentary knowledge of written German and so have to depend on translators and commentators who, by their own admission, sometimes struggle to render his complex verbal constructions and nuances into English. [93] The main reason for my hesitation, however, is Heidegger's one-time membership of the Nazi Party, which makes for a severe test of the notion that sometimes a writer's originality and importance may outweigh the unpalatable facts of their life or personality. On balance, my personal opinion is that Heidegger passes the test, and hence the section that follows. I focus solely on two of his early works, *Being and Time* and *What is Metaphysics?* (including the postscript to the latter that Heidegger added in 1943 and the lengthy introduction, 'The Way Back into the Ground of Metaphysics', he added in 1949). [94]

It is true, if conventional, to say that Heidegger marks a radical break from the tradition within Western philosophy that begins with Descartes and culminates — or exhausts itself — with Heidegger's own teacher, Edmund Husserl (to whom *Being and Time* is dedicated). Just how radical a break is suggested most tellingly by Heidegger himself when he speaks of 'prepar[ing] the transition from representational thinking to a new kind of thinking that recalls [*vom vorstellenden in das andenkende Denken*]'.[95] This is not a transition likely to appeal to most philosophers of the Cartesian tradition, who would counter that their thinking is and should be representational in the sense that it seeks to represent the world and its objects as content in the mind of an individual subject. In other words, what Heidegger is breaking away from is Descartes's whole legacy of dividing the world into 'extended stuff [*res extensa*]' or matter and 'thinking stuff [*res cogitans*]' or mind.[96] For Heidegger, Cartesian dualism represents a monumental mistake or missed opportunity in the history of philosophy because it encourages the wrong sort of questions, epistemological questions (in particular, how does content within the mind that supposedly represents the world relate to the world itself 'out there'?), and neglects or precludes the right sort of questions. And the right sort are, or rather, the biggest single question is, not epistemological but ontological: not 'What can we know about what exists?' but 'What is the nature — or "structure", as Heidegger tends to say — of what exists?' More broadly, Heidegger objects to the whole metaphysical tradition, encompassing Cartesianism and dating back as far as Plato, that deals with the world in terms of its many and various entities, or beings with a small 'b', whereas the one big question, the question of nature or 'structure', is properly about Being with a capital 'B'. That is why breaking free of Cartesian dualism also means 'overcoming metaphysics ... [and] recalling Being itself'.[97] For Heidegger, then, philosophy is a matter of recalling, not representing, and what is to be recalled is the truth of Being — the 'marvel' that what is *is*.

Who or what is capable of recalling the truth of Being, and how? In early Heidegger, at least, the answer is that it is human beings (not strictly the correct term in Heidegger's vocabulary, as we shall see, but it works well enough here) because they are special, and unique, in that their way or mode of being with a small 'b' necessarily entails an understanding of Being with a big 'B'. In language more typical of Heidegger, and we may as well start getting used to it now, this is to say that '[human beings], in their Being, comport themselves [*verhält sich*] towards their Being.'[98] For Heidegger, the way or mode of being of human beings (or of 'man', a term he does use) is what defines 'existence [*Existenz*]': 'The being that exists is man. Man alone exists. Rocks are, but they do not exist. Trees are, but they do not exist.

Horses are, but they do not exist. Angels are, but they do not exist. God is, but he does not exist.'[99]

So the understanding of Being makes the difference between 'to be' and 'to exist'. But when we ask how it is that human beings have the understanding, we come straight back to Descartes and the larger philosophical tradition. Or *not* to Descartes and the tradition, because the understanding of Being that preoccupies Heidegger could never come about if were true that we always deal with the world as subjects 'in here', separate and at a remove from objects 'out there'. While Heidegger does not deny that there are circumstances in which we do indeed treat the world in subject/object terms, he argues that these are secondary or derivative experiences. Fundamentally, he says, we are not detached and removed from the world but find ourselves — or would find ourselves if we allowed ourselves to — inseparably part of it. Dealing with the world, therefore, cannot mean turning it into so many objects for the mind to consider (with no hope of comprehending them in their entirety, 'impossible in principle'). It means coping with the world as it is, as a whole, and as we, *in its midst*, encounter it 'all the time in our experience'.

This sounds much less obscure when expressed in concrete, everyday terms. Think, for instance, of the simple action of opening a door. It is patently absurd, and just not what happens, to suggest that we first examine the door handle and consider whether it corresponds to some mental representation we have of a door handle. In reality, without thinking and maybe without even looking, we grab the handle, turn it, and open the door. Or to take another instance of what we might call 'transparent coping', a carpenter does not stop to wonder whether the hammer in their hand is as real as the hammer they have in their mind, or whether it is a hammer at all, but whacks the nail on the head and moves on to the next one, quite likely while chatting or listening to the radio at the same time.[100] Door handles and hammers are examples of entities that Heidegger terms 'ready-to-hand [*zuhanden*]' and, as a collective noun, 'equipment [*Zeug*]'. If something goes wrong (the door handle sticks, say, or the hammer proves too light to drive in a big nail), we experience such entities as 'un-ready-to-hand [*unzuhanden*]', which is one of the circumstances in which we do have to step back and treat the world in subject/object terms (because we need to diagnose and solve the problem).[101] Heidegger insists, however, it is not the primary or most characteristic way we experience the world, which remains transparent coping.

Thus far I have managed to discuss Heidegger without once mentioning '*Dasein*', probably the most important single term in his entire early vocabulary. Now is the time to rectify that.

### Dasein, Being-there, and Being-in-the-world

As indicated by the fact that Heidegger sometimes spells it '*Da-sein*', it is really two words: '*Da*' meaning 'there' and '*sein*' meaning 'being' or, for consistency within this essay, 'Being'. So Dasein translates literally as 'there-Being', but let's agree that 'Being-there' is more natural in English. Sadly, its etymology is almost the only thing about the term that can be said in a few words. Everything gets a lot more complicated when it comes to asking what Dasein is. In fact, it may be easier to start with what it is not.

Heidegger categorically denied that 'Dasein' is a synonym for 'consciousness',[102] which may well explain the disgust he expressed on encountering Jean-Paul Sartre's *Being and Nothingness*.[103] While *Being and Nothingness* is clearly based on *Being and Time*, it completely misinterprets Heidegger's intention and in effect redefines Dasein as consciousness, or the 'for-itself [*le pour-soi*]' in Sartre's terminology. Turning Dasein into consciousness (and even providing consciousness with a sort of creation story, as Sartre does implicitly in *Being and Nothingness*)[104] has exactly the consequence that Heidegger is trying to escape: it makes Dasein back into a subject in a world of objects, which perceives and represents the world to itself by sallying out from an 'inner sphere' and then 'returning with [its] booty to the "cabinet" of consciousness'.[105]

Nor is Dasein simply an entity among the other entities of the world. '*The "essence" of Dasein lies in its existence*,' Heidegger reiterates, and Dasein's characteristics are 'not "properties" ... [but] are in each case possible ways for it to be, and no more than that.... So when we designate this being with the term "Dasein", we are expressing not its "what" (as if it were a table, house or tree) but its Being.'[106] Heidegger is reminding us that ultimately the focus of his interest is not Dasein itself but its way or mode of Being — not Dasein's 'whatness' but, so to speak (and my word, not Heidegger's), its 'Beingness'.

Far from being just another entity, Dasein is 'distinguished by the fact that, in its very Being, that Being is an *issue* for it.'[107] In other words, not only does Dasein get its 'essential character' from asking the question of Being, the 'very asking' of that question *is* Dasein's mode of Being (or, strictly speaking, is 'one of the possibilities of its Being').[108] Moreover, when Dasein asks the question of Being, it already has an inkling of the answer: Heidegger calls this the '*vague average understanding of Being [durchschnittliche und vage Seinsverständnis]*'.[109] Only Dasein is capable of it. Capable of and defined by it because, according to Heidegger, 'the average understanding of Being ... in the end belongs to the essential constitution of Dasein itself' and 'It is peculiar to [Dasein] that with and through its Being, this Being is disclosed to it. Understanding of Being is itself a definite characteristic of Dasein's Being.'[110] The understanding of the greatest importance to Dasein, we would expect, is the understanding of its own (or 'ownmost')[111] Being. Sure enough, Heidegger tells us that 'Dasein always understands itself in terms of its existence — in terms of a possibility of itself....'[112] But, given that he intends *Being and Time* as an ontological project, that is, as an enquiry into the ultimate nature of Being with a capital 'B', we should not be too surprised when he goes on to claim that Dasein has an understanding of *all* modes of Being, not only its own. It is a claim he has to make, I suppose, because the ontological project would get no further if Dasein's understanding were restricted to its own Being. In any event, here is the key passage:

> Dasein also possesses — as constitutive for its understanding of existence — an understanding of the Being of all entities of a character other than its own.[113]

Whether or not we are convinced by this assertion, it does explain why for Heidegger an enquiry into the nature of all Being must start with the beings — human beings — uniquely capable of asking the question about any Being. Or to put it another way, and hijacking

Alexander Pope's more elegant wording, it explains why the proper study of Da-sein is man.[114]

Heidegger's concept of Dasein, and his lexicon of special terms associated with it, would easily fill a separate essay all of its own. For practical reasons, therefore, I'll limit myself to the five terms (or groups of closely related terms) that seem most relevant to my purpose.[115]

**Being-in-the-world.** An alternative name for Dasein is 'Being-in-the-world [*In-der-Welt-sein*]'. The preposition 'in' here has no spatial sense but serves to create a term in its own right, 'Being-in [*In-Sein*]'. Being-in denotes 'a state of Dasein's Being', not a location, and Heidegger even suggests that etymologically the original meaning of 'in' was not spatial but close to that of verbs such as 'to reside' or 'to dwell'.[116] What Heidegger is trying to get across is that Dasein is not just in the world but has a relationship with it. Not the relationship of a subject to a world of objects, needless to say, it is one that comes about through the encounter between Dasein and entities or beings other than itself — the encounter which, according to one of Heidegger's definitions, *constitutes* the world.[117] Nor is it a coincidence that Heidegger refers to Being-in's most characteristic mode of Being as 'Being-amidst [*Sein bei*]' and 'Being-already-amidst-the-world [*Schon-sein-bei-der-Welt*]'.[118] Having a relationship with the world is definitive of Dasein, not some sort of optional extra:

> Being-in is not a 'property' which Dasein sometimes has and sometimes does not have, and *without* which it could *be* just as well as it could with it.... Dasein is never 'primarily' a being which is, so to speak, free from Being-in, but which sometimes has the inclination to take up a 'relationship' towards the world. Taking up relationships towards the world is possible only *because* Dasein, as Being-in-the-world, is as it is.[119]

All of this brings us to two more of Heidegger's special terms, 'involvement [*Bewandtnis*]' and 'totality of involvements [*Bewandtnisganzheit*]'. The entities that Dasein encounters (or rather, that encounter Dasein)[120] constitute a world in the sense that they provide a background of meaningfulness — or seeming intelligibility — for Dasein's everyday activities. Think of the hammer. A hammer is a piece of equipment, defined both by what it is used for and 'in terms of its belonging to other equipment'.[121] A hammer makes sense in relation to nails, pliers and all the other paraphernalia of the carpenter's toolbox. But it also has a relation to, is 'involved' in, the carpenter's activity of hammering, and 'with hammering,' Heidegger observes, 'there is an involvement in making something fast; with making something fast, … an involvement in protection against bad weather; and this protection "is" for the sake of [*um-willen*] providing shelter for Dasein — that is to say, for the sake of a possibility of Dasein's Being.'[122] From a hammer to a house and back again, we might say, but even (or especially) in this mundane case we begin to grasp the complex totality of involvements, or 'relational totality', that forms the background against which Dasein's activities make sense. It is the background, which Heidegger calls 'significance [*Bedeutsamkeit*]', that 'makes up the structure of the world — the structure of that wherein Dasein as such already is.'[123]

**Clearing.** Yet another way of thinking of Dasein is as a 'clearing [*Lichtung*]'. Again it helps to start with Heidegger's etymology. '*Lichtung*' is literally a 'clearing' in the sense of a forest clearing, but I understand Heidegger is also punning on '*Licht*' meaning 'light'. Hence his term suggests both a metaphorical open space or openness where Dasein's encounters with the world can happen *and* the reality, according to Heidegger, that the world is encountered 'in the light of' Dasein's understanding of Being.[124] Taking the two nuances together is probably the key to unlocking the following passage:

> To say that [Dasein] is 'illuminated [*erleuchtet*]' means that *as* Being-in-the-world it is cleared [*gelichtet*] in itself, not through any other entity, but in such a way that it *is* itself the clearing.[125]

In its encounters with the world, Dasein opens or 'clears' itself to entities other than itself and it does so — is only able to do so — because it has an understanding of Being. There is, though, a twist to the story, as we are about to discover.

Heidegger never makes do with one name for a concept when he can use two, three or more, and now we have another instance. Synonymous with Being-in, the clearing also accounts for the 'there [*Da*]' of 'Being-there [*Da-sein*]'. Dasein that is 'cleared in itself', Heidegger says, '*is* in such a way as to be its "there" [*sein Da zu sein*].... By its very nature, Dasein brings its "there" along with it.' In the same passage, and just when we think we are more or less clear about the clearing, he adds, '*Dasein is its disclosedness* [*Erschlossenheit*].... [That is to say,] ... the Being which is an issue for [Dasein] in its very Being is to be its "there".'[126] In short, Being-in, clearing and disclosedness are all aspects of, or other names for, Dasein's 'thereness'. Although it is beginning to sound incoherent, I grant you, suspend your disbelief a while longer.

In simpler language, Heidegger's clearing is what we would call a 'situation', that is, the particular circumstances of any encounter between Dasein and the world. And here comes the twist in the story: every situation, Heidegger argues, is a *shared* situation. Now it is self-evidently true that some situations (attending a concert, say, or being stranded on a train that has broken down) are shared. But how is our carpenter (who, for all we know, may be hammering away on their own at a remote location miles from anywhere) in a shared situation? The answer is that their situation is shared in Heidegger's ontological sense to do with Being, not the literal sense of participating in something by being present at it. We have to look again at what Heidegger means by Dasein's understanding of Being. Before we do, we need to grapple with another of his special terms, and a particularly important one.

**The 'they'.** In most respects, Dasein seems best referred to as an 'it'. As a synonym for 'human being', however, surely it is more appropriate to speak of Dasein as a 'who', as Heidegger himself does when he asks '*who* it is that Dasein is in its everydayness'.[127] Or to ask that another way, and remembering that Dasein is not a *consciousness*, is everyday Dasein a *self*? Heidegger's reply to his own question takes us into some of the most acutely observed, and most unsettling, content of *Being and Time*:

> The self of everyday Dasein is the *they-self* [*das Man-selbst*].... As they-self, the particular Dasein has been *dispersed* [*zerstreut*] into the 'they [*das Man*]', and must first find itself.[128]

If we are flummoxed by this strange term *the 'they'*, it is fortunate that Heidegger uses it interchangeably with 'Others [*Anderen*]', which he defines very clearly (albeit accompanied by yet more new terms):

> By 'Others' we do not mean everyone one else but me — those over against whom the 'I' stands out. They are rather those from whom, for the most part, one does *not* distinguish oneself — those among whom one is too…. By reason of this *with-like* [*mithaften*] Being-in-the-world, the world is always the one that I share with Others. The world of Dasein is a *with-world* [*Mitwelt*]. Being-in is *Being-with* Others.[129]

In case we still haven't got the point, Heidegger reiterates:

> So far as Dasein *is* at all, it has Being-with-one-another as its kind of Being.[130]

The world of Dasein is the world of the 'they'. In other words, the society and culture that Dasein is with, among and in the midst of — the society and culture into which Dasein is born as a human being, and 'dispersed' as a *they-self*. Heidegger appears to be saying that Dasein is a product of conditioning, socialized into the ways and norms of a specific culture. That is what he is saying, though we should not imagine he means there is such a thing as an unsocialized Dasein that somehow then becomes socialized: by definition, Dasein is not Dasein unless it is *already* socialized into the 'they'.[131]

The 'they' cannot be identified or pinned down as any Other in particular, or even as the sum total of all Others. Alluding to his question about the 'who', Heidegger stresses that 'The "who" is not this one, not that one, not oneself [*man selbst*], not some people, and not the sum of them all. The "who" is the neuter, the "they".'[132] Paradoxically, while the 'they' is no one in particular, it holds Dasein in a tyrannical and inescapable grip. 'Dasein, as everyday Being-with-one-another,' Heidegger writes, stands in *subjection* [*Botmässigkeit*] to Others. It itself *is* not; its Being has been taken away by the Others. Dasein's everyday possibilities of Being are for the Others to dispose of as they please.[133]

Heidegger speaks of 'the real dictatorship' of the 'they' and the controlling influence of what he calls 'publicness [*Öffentlichkeit*]', for the 'they', anonymous and indeterminate itself, nevertheless determines 'who it is that Dasein is', while publicness always has the last word in defining Dasein's everydayness:

> The 'they', which is nothing definite, and which all are, though not as the sum, prescribes the kind of Being of everydayness.... Publicness primarily controls every way in which the world and Dasein get interpreted, and it is always right.... It 'was' always the 'they' who did it, and yet it can be said that it has been 'no one'.[134]

Concretely, everydayness consists in the kind of transparent, practical coping that we have already touched on. As Heidegger puts it, 'Dasein finds "itself" primarily in *what* it does, uses, expects, avoids — in those things environmentally ready-to-hand with which it is primarily *concerned*.'[135] Or more pithily, '"One *is*" what one does.'[136] Moreover, Dasein '"knows its way about" in its public environment'[137] because, as Being-with-one-another, it is 'familiar [*vertraut*]' with that environment and so comports itself in the mode of 'average everydayness [*durchschnittliche Alltäglichkeit*]' — that is, in accordance with some totality of involvements that is familiar to the 'they' within the bounds of 'averageness [*Durchschnittlichkeit*]':

> When entities are encountered, Dasein's world frees them for a totality of involvements with which the 'they' is familiar, and within the limits which have been established with the 'they's' averageness.[138]

This is the averageness not only of Dasein's average everydayness and capacity for transparent coping, but also of its average understanding of Being. Dasein has a vague sense that it already knows the answer to the question of Being because it is socialized into the averageness of the 'they' ('has grown up both into and in a traditional way of interpreting itself').[139] Dasein's familiarity with averageness *is* its understanding of Being:

> That wherein Dasein already understands itself ... is always something with which it is primordially familiar. This familiarity with the world.... goes to make up Dasein's understanding of Being.[140]

Which brings us back at last to the carpenter, whose mysteriously shared situation is now somewhat less mysterious. In Heidegger's scenario (my page 43 above), the carpenter is hammering for the sake of making a shelter against bad weather or, as Heidegger would say, the 'for-the-sake-of-which [*Worum-willen*]' of the hammering is providing shelter for Dasein.[141] But remember the qualification he added: 'that is to say, for the sake of a possibility of Dasein's Being.'[142] Heidegger may wish to stress that Dasein does not provide itself with shelter out of some instinctive urge, as a bird or a mouse builds a nest. He certainly does not describe building a shelter as a 'goal', for that would imply something that Dasein had 'in mind' as a representation or intention of its own. No, it is for the sake of a possibility of Dasein's Being. Yet Heidegger has also just told us that Dasein's world is and can only be a shared *with-world*, and that 'Dasein's everyday possibilities of Being are for the Others to dispose of as they please.' However we try to define Dasein, and we've had several goes, we cannot extricate it from the world that it shares with Others, and nor can we extricate it from the average understanding of Being that comes from, and is synonymous with, transparent coping in the shared public world. Hence the only for-the-sake-of-whichs that are available and intelligible to Dasein are the shared for-the-sake-of-whichs into which every Dasein, including every carpenter-Dasein, is socialized. To say that a carpenter does what a carpenter does may seem a meaningless tautology, but in a way it gets to the heart of the matter: what a carpenter does only makes sense in terms of *what one does* — what the 'they' does — in the particular society and culture that constitutes the with-world. In this case what a carpenter does is to hammer with a hammer.[143] And so it is that, although the carpenter may be working quite alone in the middle of nowhere, even that situation (or clearing) [144] is shared in the sense that it, too, is only intelligible — or perceived as intelligible — in the light of the understanding of Being the carpenter-Dasein shares with the 'they'.

**Care.** In Heidegger's vocabulary the term that brings together all the structural elements of Dasein's Being is 'care [*Sorge*]'. It is a synonym for disclosedness, which, as we have already seen (page 44), is itself largely synonymous with clearing:

> That by which [Dasein] is essentially cleared — in other words, that which makes it both 'open' for itself and 'bright' for itself — is what we have defined as 'care'....
> In care is grounded the full disclosedness of the 'there'.[145]

We seem to be going round in circles again. More helpfully, Heidegger once tried to convey the essence of care in a very colloquial and immediate expression: 'Being gets to me [*Sein geht mich an*],'[146] he said, meaning (as I take it) that 'care' is his name for the 'getting to me'; or as we might say, and no less colloquially, his name for the fact that 'things matter to me'. For 'me', of course, we need to read 'Dasein' here.

Predictably, Heidegger has a special word for Dasein's openness or exposedness to things mattering: '*Befindlichkeit*'. As there is no agreed translation for this word, [147] I refer to it loosely as 'receptiveness' and use it interchangeably with another of Heidegger's key terms, 'attunement [*Gestimmtheit*]' or 'mood [*Stimmung*]'.[148] It is on account of its receptiveness that things matter to Dasein in that they are, say, attractive, difficult, useful, not useful, or, as in Heidegger's own example, threatening. That the 'threatening character' of something ready-to-hand (an entity encountered by Dasein in the world: page 39) can matter is, he says, 'grounded in' Dasein's receptiveness. And as receptiveness,

> it has already disclosed the world — as something by which it can be threatened, for instance.... Dasein's openness to the world is constituted ... by the attunement of [receptiveness].[149]

Crucially, because 'we are never free of moods',[150] receptiveness as mood reveals 'the "*thrownness* [*Geworfenheit*]" of [Dasein] into its "there"':

> A being of the character of Dasein is its 'there' in such a way that, whether explicitly or not, it finds itself [*sich ... befindet*] in its thrownness. In [receptiveness] Dasein is always brought before itself, and has always found itself, not in the sense of coming across itself by perceiving itself, but in the sense of finding itself in the mood that it has.[151]

Heidegger's repeated references to Dasein's 'finding itself' reinforce the point that he is making: not only is Dasein Being-in-the-world, in its moods — 'not of its own accord' — it finds itself *thrown into* the world and into its 'there'.[152]

Dasein, it appears, is completely at the mercy of things and events in the world, passive and impotent — a very sorry sort of Being, perhaps a lost cause. It is worth saying now, therefore, that this only tells us half the story about Dasein, because there is a tension between thrownness and another of Dasein's essential characteristics, 'projection [*Entwurf*]':

> Dasein has a kind of Being in which it is brought before itself and becomes disclosed to itself in its thrownness. But thrownness, as a kind of Being, belongs to a being which in each case *is* its possibilities, and is them in such a way that it understands itself in these possibilities and in terms of them, projecting itself upon them.[153]

Heidegger identifies projection with the structure of Dasein's understanding:

> As projecting, understanding is the kind of Being of Dasein in which it *is* its possibilities as possibilities.[154]

As transparent coping is directed towards for-the-sake-of-whichs, not ideas or intentions that Dasein has 'in mind', it comes as no surprise to hear that projection has nothing to do with 'goals' or with

> comporting oneself towards a plan that has been thought out, and in accordance with which Dasein arranges its Being. On the contrary, any Dasein has, as Dasein, already projected itself; and as long as it is, it is projecting.[155]

For Dasein, projecting itself does not mean working to a plan but *pressing forward into* possibilities and acting upon them. Hence Dasein is 'essentially ahead of itself' because it is always pressing towards the future, towards understanding itself in and by some new possibility in the future:

> Dasein is already *ahead* of itself in its Being. Dasein is always 'beyond itself', not as a way of behaving towards other entities which it is *not*, but as Being towards the potentiality-for-Being [*als Sein zum Seinkönnen*] which it is itself.[156]

If, as Heidegger has told us, Dasein's everyday possibilities are for Others to dispose of as they please, it follows that the only possibilities available to Dasein are those determined by the 'they' — the society and culture into which Dasein is born and 'dispersed'.[157] Dasein is both thrown and projected, a 'thrown projection'.[158] But how does that change anything? Dasein is still at the mercy of entities and events, and even if possibility A happens to open up for it, possibilities B, C and D may be closed off as they do not accord with the publicness and averagenesss of the 'they'. And yet Heidegger suggests — ambiguously and not altogether convincingly, it has to be said — that there is potential here for something to change because, in the tension between Dasein's thrownness and its capacity to project itself into possibilities, there is scope for it to make a choice. More anon about this.

**Anxiety.** Of all the moods to which Dasein is receptive, for Heidegger the most fundamental is 'anxiety [*Angst*]'. Not to be confused with 'fear', which is fear of something or someone in particular, anxiety is a diffuse sense of unease that cannot be located or attributed to any one source. 'That in the face of which one is anxious,' Heidegger insists, 'is completely indefinite':

> [It] is characterized by the fact that what threatens is *nowhere*.... [but] is already 'there', and yet nowhere.... The obstinacy of the 'nothing and nowhere within-the-world' means as a phenomenon that *the world as such is that in the face of which one has anxiety*.[159]

Which is to say:

> That in the face of which one has anxiety [*das Wovor der Angst*] is Being-in-the-world as such.[160]

A few pages later Heidegger spells this out more precisely:

> anxiousness as [receptiveness] is a way of Being-in-the-world; that in the face of which we have anxiety is thrown Being-in-the-world; that which we have anxiety about is our potentiality-for-Being-in-the-world [*In-der-Welt-sein-können*].[161]

Along the way, Heidegger has introduced an important new term to describe anxiety, '*unheimlich*'. It is usually translated as 'uncanny' or 'unsettled', but has the more literal meaning — and the one intended by Heidegger — 'unhomelike' or 'not at home':

> In anxiety one feels 'unsettled'. Here the peculiar indefiniteness of that amidst which Dasein finds itself in anxiety comes primarily to [be expressed]: the 'nothing and nowhere'. But here 'unsettledness' also means 'not-being-at-home [*Nicht-zuhause-sein*]'.[162]

In *What is Metaphysics?*, first published two years after *Being and Time*, '*unheimlich*' takes on other, even more perplexing nuances:

> What is 'it' that makes 'one' feel ill at ease? We cannot say.... All things and we ourselves sink into indifference.... We can get no hold on things. In the slipping away of beings only this 'no hold on things' comes over us and remains.
>
> Anxiety reveals the nothing [*Die Angst offenbart das Nichts*].
>
> In anxiety we are 'in suspense'. More precisely, anxiety leaves us hanging because it induces the slipping away of beings as a whole.... In the altogether unsettling experience of this suspendedness where there is nothing to hold onto, pure Da-sein is all that is still there.[163]

Anxiety reveals the nothing. Or as Heidegger rephrases it, 'Da-sein means being *held out into* the nothing [*Hineingehaltenheit in das Nichts*].'[164] So we have another question: what is this '*Nichts*', this 'nothing'?

The answer seems to be that it is the true nature of Dasein, which is a nothing (or 'nullity [*Nichtigkeit*]') in the sense that it is *nothing but* the possibilities, the averageness, and the for-the-sake-of-whichs of the 'they'. In other words, anxiety reveals to Dasein what Heidegger has been telling us, his readers, for some time: that the possibilities Dasein presses forward into, and the way it understands itself, are determined — dictated — by publicness and Others, not by Dasein itself. Anxiety confronts Dasein with its illusions and reveals the reality that it has no meaningful basis or grounding of its own, and no hope of acquiring one.

Evidence to support this interpretation is scattered throughout *Being and Time*. At one point, for example, Heidegger explicitly declares that Dasein 'is not itself the basis of its Being', but '*is* a nullity of itself.'[165] More indirectly, he says that anxiety 'brings [Dasein] back from its absorption in the "world"' with the result that '[e]veryday familiarity collapses':

> 'The world' can offer nothing more, and neither can the Dasein-with of Others. Anxiety thus takes away from Dasein the possibility of understanding itself ... in terms of 'the world' and the way things have been publicly interpreted.[166]

Heidegger returns to this theme later, keen to stress that anxiety 'discloses an insignificance of the world'.[167] That is, anxiety deprives Dasein of the background of involvements ('significance') that normally provides the structure of its world and makes sense of its activities. Thus deprived, Dasein 'clutches at the "nothing" of the world'. In anxiety Dasein *disappears* and all that remains is the anxiety itself, which is why Heidegger suggests it is anxiety, not Dasein, that is 'anxious in the face of the "nothing" of the world'.

**Two new questions, one straight answer, and one not so straight**

If I have spent a disproportionate amount of time on Being-in-the-world, clearing, the 'they', care and anxiety, it is because I think we have to decipher these terms in particular before we can get to the heart of what Heidegger is saying. And before we can answer two more questions, big questions to which all the others have been leading. Firstly, has Heidegger succeeded in convincing us that thinking that 'recalls' is more truthful and meaningful than thinking that 'represents'? Secondly, and if he has succeeded, does he tell us what practical difference it could or should make to how we live our lives, individually and perhaps collectively?

My own answer to the first question is brief and comes down to the proposition that, once read, *Being and Time* and *What is Metaphysics?* cannot be unread. In breaking free from the Western philosophical tradition, early Heidegger stands Descartes on his head: no longer thinking stuff, conscious subjects in a world of objects, we just find ourselves in the world — thrown there, not of our own accord, Being-already-amidst-the-world. Heidegger has shifted the burden of proof, as it were, and what we must explain and justify now is not why we should stay with Heidegger's way of seeing the world, but why we should go back to seeing the world exclusively, or even primarily, in terms of subjects and objects. Heidegger is not suggesting for one moment that there is no such thing as subject/object thinking: on the contrary, in his own terms he acknowledges that without it not only would solutions to mundane problems (the door handle that sticks, the hammer that fails to do its job) be impossible, so would scientific theory.[168] But he is suggesting, convincingly in my opinion, that subject/object thinking fails to account for our commonest and most characteristic experience of the world: that kind of experience, he says, consists in transparent coping against a background of shared understanding of what makes sense in our culture and society.

Answering the second question takes us from the burden of proof to the burden of expectations. Descartes starts with the individual, self-contained subject (or '"cabinet" of consciousness', as Heidegger mocks) and then considers its relations with the world. Heidegger reverses the sequence and makes the world (as he defines it) his starting point. In so doing, Heidegger both dispels what he regards as the Cartesian illusion and sets himself an enormous challenge: not Descartes' challenge of explaining how an individual consciousness 'in here' can relate to the world 'out there', but the exactly opposite challenge of explaining how, even in principle, anything individual at all (let alone self-contained) could ever be possible in the dictatorial with-world of publicness and the 'they'. As I forewarned a while ago, Heidegger's response to this challenge is convoluted and ambiguous, which means that my answer to the second question is likewise somewhat convoluted and ambiguous.

The key passage in *Being and Time* is one I have already alluded to (my page 41). In full it reads:

> Dasein always understands itself in terms of its existence — in terms of a possibility of itself: to be itself or not itself. Dasein has either chosen [*gewählt*] these possibilities itself, or got itself [*hineingeraten*] into them, or grown up [*aufgewachsen*] in them already.[169]

The word that leaps out is 'chosen'. Why on earth does Heidegger speak of choices and choosing when he has been at such pains to impress upon us that Dasein is in the grip of the 'they', powerless to determine anything for itself? We need to look again at what Heidegger has just said. In understanding itself 'in terms of a possibility of itself', it emerges, Dasein has some scope to choose: *to choose these possibilities itself*, as opposed to *getting into them*, or finding that it has already *grown up in them*. Colloquially and almost casually, Heidegger is telling us where to look for the answer to the question of how we could or should live differently. To look, that is, to the tension between Dasein's thrownness and its capacity to project itself into possibilities, which presents Dasein with three alternative modes of existence. They are, less colloquially, the 'authentic [*eigentlich*]' mode; the 'inauthentic [*uneigentlich*]' mode; and that in which 'neither ... [authenticity nor inauthenticity] has been differentiated [*die modale Indifferenz ihrer*]', or the 'undifferentiated' mode for short.[170] Not quite in that order, let's take each of them in turn.

**Dasein has chosen the possibilities itself**. Here 'chosen' is shorthand both for 'chosen authentically' and for 'chosen in the face of anxiety', as is clear from the following passage:

> Anxiety makes manifest in Dasein its *Being towards* its ownmost potentiality-for-Being — that is, its *Being-free for* the freedom of choosing itself and taking hold of itself. Anxiety brings Dasein face to face with its *Being-free for* ... the authenticity of its Being, and for this authenticity as a possibility which it always is.[171]

The weight of expectations grows, for now Heidegger is saying that Dasein has 'freedom [*Freiheit*]' — and freedom not only to choose but also to 'take hold' of itself. Despite all appearances to the contrary, Dasein does have the prospect of escaping the grip of the 'they', and what had seemed to be a philosophy of impotent conformity has turned out to be a philosophy of personal liberation.

Or has it?

The liberation Heidegger offers does not release Dasein from the tyranny of publicness and the 'they'. We were right about that the first time. Dasein is still not free to create its own possibilities from scratch, only to press into possibilities made available by the culture and society in which it dwells. 'Freedom' does not signify liberation from publicness and Others, but from everyday Dasein's illusion that it has some identity and meaningful grounding of its own, or any hope of acquiring one. This will sound familiar because it is the same illusion that Dasein is confronted with by anxiety (my page 52): hence 'freedom' can also be defined as 'anxiety [that] is *held on to* when one brings oneself back to one's ownmost thrownness', which is to say, when 'Dasein is taken all the way back to its naked unsettledness [*nackte Unheimlichkeit*], and becomes dazed [*benommen*] by it.'[172] Even holding on to anxiety, however, is not enough on its own to make Dasein capable of authenticity.[173] There must be a transformation, a moment in which Dasein is finally open to its own groundlessness, its suspendedness in the nothing, and 'takes over authentically in its existence the.... nullity by which [its] Being is dominated primordially through and through'.[174] Heidegger's term for the moment of transformation is the '*Augenblick* [literally, *glance of an eye]*.[175] His term for what results from the transformation is '*Entschlossenheit*', which is translated as 'resoluteness' but, read as '*Ent-schlossenheit*', I understand works as a pun suggesting 'un-closedness' or 'openness'.

Transformed and therefore authentic at last, for Dasein everything changes — and, a seeming paradox, nothing changes. Nothing changes in the sense that, as we now know for sure, Dasein cannot escape publicness and the shared, average everydayness into which it is socialized. Yet authenticity does not demand that Dasein should escape its socialization: 'authentic existence,' Heidegger says, is not something that 'floats above' everydayness, *only a modified way in which ... everydayness is seized upon* [*nur ein modifiziertes Ergreifen dieser*].'[176] In other words, for Heidegger authenticity is defined not by what Dasein does, but by how Dasein interprets to itself — or, his phrase, 'seizes upon'[177] — what it does. By that definition, everything changes for a resolute, authentic Dasein.

It would be too much to expect Heidegger to have set out the changes in a few handy bullet points. Since he does not, the following is my own attempt, which, while speculative in places, hopefully captures the major consequences of Dasein's transformation to authenticity.

- Resolute Dasein no longer responds to the 'general situation [*allgemeine Lage*]' but to the 'Situation [*Situation*]'.[178] Although it continues to press into possibilities that may be available to other Daseins (one of the nuances of 'general situation', I assume), it does so in a way that is *unique to itself*: openly, spontaneously, and free of the illusion that those possibilities themselves offer it any grounding, or embody anything other than averageness and publicness.

- Resolute Dasein has an individual identity distinct from the they-self of everyday Dasein, an '*authentic self* ... which has been taken hold of in its own way [*eigens ergriffenen*].'[179] Another apparent paradox, this is possible because of, not in spite of, Dasein's recognition that no identity or grounding is to be found within the possibilities it shares with other Daseins.

- Authentically 'taking action' is not a matter of weighing up alternative possibilities and deciding between them (as a subject in a world of objects might do). Dasein's openness to the Situation and its authentic 'action' are simultaneous, if not synonymous:

  > Resoluteness does not first take cognizance of a Situation and put that Situation before itself; it has put itself into that Situation already. As resolute, Dasein is already *taking action*.[180]

- In the moment of transformation, Dasein ('Being-there') is not just 'there' but 'more authentically "there" … as regards the Situation which has been disclosed.'[181] The quality of Dasein's 'thereness' is itself transformed.

- Unlike an everyday Dasein (so paralyzed by anxiety that it disappears and leaves only anxiety itself anxious before 'the "nothing" of the world'), an authentic Dasein is not 'inhibited' or 'bewildered' by anxiety even as it holds on to it:

  > He [sic] who is resolute knows no fear; but he understands the possibility of anxiety as the possibility of the very mood which neither inhibits nor bewilders him.[182]

Resolute Dasein, then, is responsive to the Situation, not the general situation; is an authentic self, not a they-self; is no sooner in a Situation than 'acting' authentically and 'there' more authentically; and, far from being overcome by anxiety, faces anxiety with equanimity.[183] There is one more surprise.

- Authenticity brings joy to Dasein:

  > Along with the sober anxiety which brings us face to face with our individualized potentiality-for-Being, there goes an unshakable joy [*gerüstete Freude*]….[184]

In Heidegger's terms, and especially by contrast with all he has said about everyday Dasein, we cannot deny that Dasein's authentic transformation qualifies as a form of personal liberation. Moreover, the *joyful* freedom it offers sounds a lot more appealing than the *absurd* freedom offered in *Being and Nothingness*, Sartre's reworking of *Being and Time*, which tells us that we are 'condemned to be free', must make our own meanings and values, and do so 'without justification and without excuse'.[185]

On the other hand, Heidegger's liberation is a very austere affair, abstract and unadorned as well as demanding, with no obvious, practical implications for how we might live differently even at the individual level, let alone collectively. To my knowledge, Heidegger nowhere suggests that we renounce society altogether and subsist hermit-like in the mountains. Quite the reverse: living authentically looks pretty much like business as usual as far as our day-to-day circumstances are concerned. All that is different is that we now interpret our circumstances, ourselves, and the world with a sort of ironic detachment, no longer expecting a meaningful grounding for our lives and yet, Heidegger would have us believe, rejoicing in our freedom from such expectations. That is, although Heidegger does not urge us literally to lead the life of a hermit, his liberation itself seems forbiddingly severe in nature, and perhaps not a great improvement on Sartre's absurd liberation after all.

**Dasein has already grown up in the possibilities.** Authenticity is very difficult to achieve and sustain precisely because the 'self' of everyday Dasein is 'primarily and usually inauthentic, the they-self. Being-in-the-world is always fallen.'[186] Confusingly, when Heidegger says 'inauthentic' here he probably means 'undifferentiated', for the context indicates that he is referring to the mode (the third, not the second, of the three identified on my page 54) in which Dasein is neither authentic nor inauthentic.[187] Assuming that is what he means, how are we to take 'Being-in-the-world is always fallen', which implies that the undifferentiated mode is necessarily a 'fallen' mode?

In Heidegger's lexicon 'fallenness [*Verfallenheit*]' is a key term in its own right and one of three components of care.[188] We also find 'falling [*Verfallen*]' as a noun but because he uses it in several different ways its relation to 'fallenness' is not always clear.[189] I shall avoid 'falling', therefore, and focus specifically on the intriguing expression 'always fallen'. It translates more literally as 'always already fallen [*immer schon verfallen*]' and conveys, like the biblical story of the Fall, a sense that there is something inherently wrong with man. Rather cryptically, but surely an explicit allusion to the Fall, Heidegger declares that 'Dasein prepares for itself a constant temptation towards falling. Being-in-the-world is in itself *tempting* [*versucherisch*].' [190] What is wrong and tempting, as I understand Heidegger, has nothing to do with the disobedience that got man banished from Eden, and everything to do with the anaesthetizing or 'tranquillizing [*beruhigend*]' effect of Dasein's socialization. In Heidegger's words:

> [T]he everyday publicness of the 'they' ... brings tranquillized self-assurance — being-at-home [*Zuhausesein*]' ... — into the average everydayness of Dasein. [191]

The result of tranquillization is that Dasein, 'lost' in the 'they' and 'carried along by the nobody', [192] is relieved of the unsettledness of *not*-being-at-home. In other words, and crucially, tranquillization enables Dasein to flee the anxiety that reveals its unsettledness. By the same token, tranquillization helps to explain why Dasein finds itself sharing in the illusion (which is, of course, the 'they's' own illusion) that publicness and averageness provide a meaningful grounding for life:

> The supposition of the 'they' that one is leading and sustaining a full and genuine 'life' brings Dasein a *tranquillity* [*Beruhigung*], for which everything is 'in the best order' and all doors are open. [193]

The reason these things happen is that Dasein's commonest and most characteristic mode of existence — its default mode — is the undifferentiated, in which Dasein is *always already fallen* because it has *grown up in* possibilities made available by the 'they'.

**Dasein has got itself into the possibilities.** The missing piece in the jigsaw (i.e., the second of Dasein's three modes of existence) is the inauthentic. A clue to telling it apart from the undifferentiated mode, which it closely resembles, may be this curious sentence:

> Everydayness is determinative for Dasein even when it has not chosen the 'they' for its 'hero [*Helden*]'. [194]

By indicating that sometimes Dasein *does* take the 'they' for its 'hero', Heidegger seems to be saying that we have to distinguish between the undifferentiated kind of inauthenticity, which is passive and unwitting (to the extent that it results solely from Dasein's socialization), and another kind, 'the "they" as hero' inauthenticity, in which Dasein actively and wittingly identifies with the 'they' (or more correctly, with possibilities and for-the-sake-of-whichs determined by the 'they'). The latter kind, we might say, clearly constitutes inauthenticity, whereas the nature of undifferentiated inauthenticity remains ambiguous.

I am all too aware that I have still not answered, or not fully answered, my own second question: what are the practical implications for how we live, if any, of thinking that 'recalls' as opposed to thinking that 'represents'? To finish the job I'll try once again, and for one last time, to decipher Heidegger's language.

There is a passage in *Being and Time* in which he characterizes authenticity, the undifferentiated mode, and inauthenticity, in that order and in these terms:

> [Because] Dasein is in each case essentially its own possibility, it *can*, in its very Being, 'choose' itself and win itself; it can also lose itself and never win itself; or only 'seem' to do so.[195]

*Not* in the same order, the following is my understanding of what Heidegger means, and how it relates to the question I am attempting to answer.

**Authenticity.** In resolute authenticity, we 'win' ourselves in the sense that we are free of the illusion there is or can be any grounding to our lives, and thereby 'take hold of' an individual identity distinct from the they-self. More authentically 'there' in the world, we face anxiety not with unsettledness but with 'unshakable joy'. According to Heidegger, none of this demands fundamental change in the day-to-day circumstances of our lives. Even though it may make no practical difference to how we live and behave, however, authenticity does demand that we 'seize upon' (i.e., interpret) our everydayness in a way that liberates us from tranquillized conformity with publicness and averageness. To repeat what I said earlier, the problem is that this demand would test the resolve of a hardened ascetic, never mind the average reader of *Being and Time*. In short, it is extremely difficult to imagine anyone actually achieving Heidegger's sort of authenticity, or sustaining it for long if they did.

**Inauthenticity.** I take the liberty of expanding the elliptical ending of Heidegger's sentence from 'or only "seem" to do so' to 'or only "seem" to win itself while in fact forgoing or disowning the opportunity to take hold of an authentic self.'[196] What Heidegger is describing is 'the "they" as hero' inauthenticity, Dasein actively identifying with one or more of the for-the-sake-of-whichs made available by the society in which it happens to dwell. So far, so bad. But is it really so bad? Does Heidegger mean 'inauthenticity' as a term of denigration or reproach? It appears not, or not necessarily. After all, in any society or culture there will be a vast array of for-the-sake-of-whichs, encompassing every conceivable social role or social identity, profession, occupation or endeavour, including many (parent, doctor, nurse, teacher, rescue worker, peace negotiator, to name but a few obvious and uncontroversial examples) that are especially respected and valued, and presumably Heidegger himself would have respected and valued. Despite the negative-sounding connotations of the word, therefore, he appears to mean 'inauthenticity' in a technical and non-judgemental sense. By his definition of authenticity, it is true, even the most devoted and loving of parents, the most skilful of doctors, the most inspirational of teachers, and so on, only seem to 'win' themselves without actually doing so. Yet as Heidegger would agree, I think, there is no denying the value of their lives within the constraints of everydayness.

**The undifferentiated mode.** In the case of the undifferentiated mode, which is barely distinguishable from the inauthentic in most respects, Heidegger makes it clearer still that these are not pejorative terms. On the contrary, and as if to forestall any such misunderstanding, early in *Being and Time* he stresses that the 'undifferentiated character of Dasein's everydayness is ... a positive phenomenal characteristic.'[197] He ascribes a positive, not a negative, quality to our default mode of everyday existence. What are we to call this quality? Not 'authenticity', of course, but how about something like 'genuineness'? Or 'everyday genuineness'? It is difficult to know what to call it, which is symptomatic of the larger problem that Heidegger's terminology tends to give rise to confusing or contradictory messages. And that, in turn, makes it all the harder to tell whether he seriously expects *Being and Time* to change how anyone lives or behaves. If we can live full and worthwhile lives (albeit inauthentically, according to Heidegger) by actively identifying with the possibilities available to us, and if even our default mode of existence (the undifferentiated, likewise inauthentic in all but name) is 'positive', what reason or incentive do we have to attempt something — authenticity, 'winning' ourselves — that by Heidegger's own account sounds virtually impossible in the first place?

# 6. Nothingness and suchness: the Sixth Patriarch, *śūnyatā*, and Nishida's notion of 'the good'

## The Sixth Patriarch

The variety of Buddhism that is generally known by its Japanese name, Zen, is called Chan in Chinese. It is thought to have been brought to China from India in the fifth or early sixth century AD by a monk named Bodhidharma, later designated as the First Patriarch of Chan. Time passed and one patriarch succeeded another. The Fifth Patriarch was Hongren, who resided at a monastery in what is now the province of Hubei in central China. He is believed to have died in 674. When Hongren was looking to choose his own successor, or so the story goes, he invited his disciples to compose a poem that would demonstrate their suitability. Only one monk responded — Shenxiu, who, anonymously and in the dead of night, wrote the following verse on a wall:

> Let the body be a *bodhi* tree,
> The mind a shining mirror on a stand;
> Take care to keep the mirror clean,
> Lest it attract the dust.[198]

It was assumed that Hongren's mantle would pass to Shenxiu, and publicly Hongren himself seemed to favour this outcome. Judging by the poem, however, which he knew had been written by Shenxiu, Hongren in fact doubted that Shenxiu had attained true enlightenment. Then events took an unexpected turn. On hearing Shenxiu's poem recited aloud, Huineng, who was illiterate and performed menial tasks in the monastery, composed one of his own and asked for it to be written for him on the wall beside Shenxiu's. It read:

> There never was a *bodhi* tree,
> Nor shining mirror on its stand;
> There never was a single thing,
> Whence then should come the dust?

Huineng's poem was seen by Hongren, who secretly visited Huineng while he was hulling rice and summoned him to a private audience very early the next morning. There Hongren satisfied himself that, unlike Shenxiu, Huineng had attained enlightenment and appointed him his successor. Hongren was well aware that once his choice was revealed it would be bitterly resented by many within the monastery, which is why he advised Huineng to leave at once. So it was that Huineng fled south and in time came to be identified with the tradition known as Southern Chan. With pleasing symmetry, Shenxiu went north and later became identified with Northern Chan.

The story is almost certainly apocryphal, put together long after the events it purports to record.[199] Conventionally it is interpreted as characterizing the differences between the so-called 'gradual' school of Chan (synonymous with Northern Chan) and the 'sudden' school (Southern Chan). Shenxiu's poem stands for an exacting, step-by-step progression towards enlightenment, one which demands the renunciation of all worldly attachments (the dust that might obscure the mirror), meditation, close study of sacred texts — in short, a life of rigorous and time-consuming spiritual discipline. Huineng's riposte, by contrast, expresses enlightenment not as a gradual progression towards some state of being in the future but as a spontaneous intuition that *we have been in that state all along*, if only we realized it. As symbolized by Huineng's illiteracy, this sort of enlightenment cannot be conveyed by the written word (a metaphor for methodical, rational understanding) and, as symbolized by his lowly standing in the monastery, it is not the preserve of those with rank and status.

While that is the conventional interpretation, surely the story is not only about the merits of one religious sect versus another. Far more profoundly, it is about the nature of reality, and how we conceive of it — or fail to conceive of it.

### *Śūnyatā*

For Buddhism, reality is '*śūnyatā*', the great nothingness (or emptiness) that is without time, space or other differentiation and yet gives rise to the world we know: time, space and the minutest differentiations of everything, animate and inanimate, that is and might ever be.[200] Huineng's 'There never was a single thing' does not deny that the world and its things exist, only that they exist 'singly', that is, are somehow 'outside' or 'apart from' *śūnyatā*. Everything in the world is both uniquely differentiated in itself and inseparably part of the undifferentiated nothingness. As if all this were not paradoxical and baffling enough, everything that has come out of the nothingness exists with a quality of distinctness and immediacy for which we barely have a word in English. Buddhists refer to it as *suchness* — the suchness of a particular plant, for instance, or a particular creature, or a particular rock, *just as it is* and not otherwise. [201] Buddhist enlightenment consists in the dissolution of all the cares, preoccupations, ambitions, prejudices, illusions, delusions, and myriad other barriers that stand between us and the suchness, the just-as-it-isness, of the world and of ourselves. It consists, in other words, in an unmediated communion with reality.

The Buddhist conception of reality seems strange and unsettling to anyone more familiar with the Judaeo-Christian tradition. One way of measuring how strange and unsettling is to remind ourselves of the state of knowledge that had enabled Adam to name the beasts of the field, 'giv[ing] unto every living creature a name according to his propriety', as Bacon puts it (or 'a suitable name that harmonizes with its nature', as Luther puts it). What 'gave occasion to the fall,' Bacon insists, was not that 'pure light of natural knowledge' but

> an aspiring desire to attain to that part of moral knowledge which defineth of good and evil, whereby to dispute God's commandments and not to depend upon the revelation of his will, which was the original temptation. [202]

On one interpretation, Bacon implies something akin to the Buddhist notion of suchness. He agrees that it was not for Adam to decide whether the animals were good, bad or indifferent. That was God's job. At the same time, though, isn't Bacon saying that it *was* for Adam — was, indeed, the proper exercise of his 'natural' knowledge — to see and name the animals in all their distinctness and intensity, from the aardvark in its aardvarkness to the zebra in its zebraness? Yes and no. Although Bacon probably would recognize the suchness of the animals, for him it would (like the animals themselves) be the work of God, a divinity 'above' and 'beyond' everything he has brought into being, creator to his creation, subject to his object, a 'there' to the 'here' of our world. And at this point the whole comparison breaks down, because the conception of reality to be found in Christianity is ultimately incompatible with the Buddhist conception of reality. Where Christianity conceives of reality fundamentally in terms of *duality*, beginning with God on one side and the world and man on the other, Buddhism denies any kind of duality: not just the duality of a 'natural' knowledge appropriate to man and a 'moral' knowledge appropriate to God alone, but the duality of knowledge and knower, subject and object, a *bodhi* tree as opposed to an oak tree, a clean mirror as opposed to a dusty mirror. Above all, crucially and inescapably, Buddhism denies the duality of creation and a creator, of man and a God. When Milton and Bacon speak of 'regaining to know God aright' and recovering 'the pure light of natural knowledge' they express their devout hope for a spiritual reunion with God. Yet they know that the reunion can never be complete, that the best we can expect is to 'be like [God], as we may the nearest' because, even in principle, the gulf between the Godness of God and the manness of man is unbridgeable. They are two different orders of reality. But in Buddhism there is only one order of reality, the great nothingness, *śūnyatā*. This is what Huineng expresses when he asks his rhetorical and seemingly banal question, 'Whence then should come the dust?' The dust exists in the sense of its suchness; it does not and could not exist 'singly' in any sense that would imply 'separate from' *śūnyatā*. For Milton and Bacon, reality consists in an incomplete reunion with God. For Huineng, reality consists in — reality *is* — the intuition that there is no disunion, no duality, to be bridged.

Spare a thought for Huineng, who must be shaking his head in bemusement. He has just told us, after all, that the written word is of little value as a means to apprehend the nature of reality, and yet here we are carrying on regardless, groping for the right words — or any words — to grasp *śūnyatā* as a description of reality. The conundrum is that, except for those who practise Chan/Zen in meditation or other essentially non-verbal forms (in Japan, typically tea ceremony, flower arranging, art and calligraphy, garden design, archery), words are all that most of us have to go on. For better or worse, we are a verbal species and unless the nature of reality is somehow communicable in words, however imperfectly, it will remain an esoteric secret known to the few, not the many, or any significant proportion of the many. Someone needs to demonstrate that Huineng's aversion to the written word is overzealous, and self-defeating because it threatens to leave reality hidden in the monastery, if not some holy hermit's cave. Who would be bold, or foolhardy, enough to attempt the demonstration? Well, one man who did was the Japanese philosopher Nishida Kitarō.

## Nishida Kitarō, warts and all

By common assent, Nishida was Japan's most important twentieth-century philosopher and the originator, in effect, of a whole school of philosophy, the so-called Kyoto School.[203] Born in 1870 in Ishikawa Prefecture, he grew up in the early years of the Meiji Period, when Japan was opening to the world after more than two centuries of self-imposed isolation. In every conceivable way, including the assimilation of foreign ideas as well as technologies, it was a time of profound change and re-evaluation in Japan. Despite dropping out of his secondary school, Nishida managed to get himself accepted into the philosophy department of Tokyo Imperial University, graduating in 1894. After years spent teaching German, philosophy and other subjects in provincial schools, he was appointed assistant professor of philosophy at Kyoto Imperial University in 1910, and then a full professor there in 1913. He retired in 1928 but continued lecturing and writing right up until his death in 1945.

Soon after graduation, Nishida had begun Zen meditation, which he practised intensely from 1897. For unknown reasons, he gave up meditation in 1905 and, except in his last writings, rarely makes any explicit reference to Zen. Nevertheless, Zen is very much part of what I wish to say about Nishida.

It would be wrong to end this biographical sketch without noting that Nishida's later years are overshadowed by controversy. Wittingly or unwittingly, he strayed into political philosophy and allowed himself to be drawn into an ideological debate in Japan dominated, before and during the Pacific War, by nationalists and militarists. Some of his utterances, particularly on the 'national polity [kokutai]', 'Japanese spirit' and the supposedly unique significance of the emperor system, were interpreted as endorsing Japan's aggressive expansion in Asia — interpreted, that is, as a perverse philosophical justification for the Greater East Asia Co-Prosperity Sphere.[204] As with Heidegger, so perhaps with Nishida: the disappointment of finding that philosophers have feet of clay.

It also has to be said that Nishida is not an easy read. One of the difficulties is that he is often engaged in a sort of running dialogue with Western philosophers, whose ideas he variously explores, adapts or refutes. They include not just major figures such as Plato, Aristotle, Hegel and Kant [205] but a number of philosophers, especially Heinrich Rickert and other German neo-Kantians, who, though influential in Nishida's day, are of more limited interest now. The main obstacle to understanding Nishida, however, is his own writing style. Tortuous in wording and structure (or lack of structure), repetitive and, for page after page, numbingly abstract, Nishida's prose comes across as a relentless stream of ideas committed to paper with scant consideration for anyone else trying to follow the train of thought.[206] The most memorable passages are those in which Nishida allows himself a metaphor or analogy. For instance, touching on the relation between the universal and the particular (or the one and the many, as we might say), he invokes a musical analogy. Our conventional sense that individual sounds combine to make a melody is mistaken, he suggests: on the contrary, the individual sounds are *already there* in the melody, '*placed* [*oite aru*]' within it.[207] Elsewhere Nishida makes the same point with an image from calligraphy or ink painting:

> True universality and individuality are not opposed to each other.... Each of the artist's exquisite brush strokes expresses the true meaning of the whole.[208]

Ironically, the very memorableness of these brief moments of figurative language, contrasted with the unmemorableness of the bulk of Nishida's prose, leaves us wondering whether he shouldn't have heeded Huineng's warning that the written word is of little help in getting to the true nature of reality. So why, you are bound to ask, should we even attempt to read Nishida? My answer concerns the spirit, if not the letter, of Nishida's philosophical project. This will take a bit of explaining.

## Pure experience

Nishida's first major work, published in 1911, was titled *Zen no kenkyū*, which translates as *An Inquiry into the Good*.[209] It is the most accessible of his writings, in my opinion, and the one I quote from most frequently. The opening lines of the first chapter, 'Pure Experience', read as follows:

> To experience means to know facts just as they are [*jijitsu sono mama ni shiru*].... What we usually refer to as experience is adulterated with some sort of thought, so by *pure* I am referring to the state of experience just as it is without the least addition of deliberative discrimination.[210]

Nishida's final work, completed only months before he died in June 1945, was '*Bashoteki ronri to shūkyōteki sekaikan*', which translates as 'The Logic of *Basho* [or *Topos*][211] and the Religious Worldview'. Towards the end of it is this passage, awkwardly worded but striking in its own way:

> What is both nearest and furthest away is what is most true. No matter how far we may go, truth arises not in the loss of its [truth's] starting point but, on the contrary, in the return to it. This is what I call 'acting-intuition [*kōiteki chokkan*]'.[212]

From one perspective, these two quotations trace the arc of the development of Nishida's philosophy, culminating in the notion of 'acting-intuition' (his term for what he sees as human beings' dynamic, two-way relationship with the world).[213] From another perspective, however, they trace a circle rather than an arc, for Nishida's philosophy itself sets out on a long journey only to return — and return to that first chapter of *An Inquiry into the Good*.

Nishida's definition of pure experience seems unambiguous.[214] What, though, are the 'facts [*jijitsu*]' that pure experience knows? The answer is that they are the facts of reality, 'true' reality, which is why I suggest that Nishida's first sentence could and should also be read:

> To experience means to know reality just as it is....

By Nishida's own account, the idea of 'actuality just as it is [*genjitsu sono mama no mono*]'[215] was one he had had from childhood. In any case, true reality — known through pure experience — is the starting point from which Nishida's philosophy departs, and to which it returns.

The sort of thought that 'adulterates' experience, Nishida soon makes clear, is thought that seeks to know the world by dividing it into 'knowing and its object'.[216] It is what here he calls 'deliberative discrimination [*shiryo bunbetsu*]' and in later works 'object logic [*taishō ronri*]', that is, a logic (or, shall we say, frame or habit of mind) that treats everything in terms of subject and object, or some other form of dualism. The conventional distinction between the universal and the particular is another such dualism, as we have already seen. But for Nishida experience is 'adulterated' by any and every kind of dualistic thinking whatsoever, including thinking that assumes *a duality of being as opposed to nothing, and of the individual self as opposed to the universe as a whole.*

What enables direct, unadulterated, pure experience is, in the language of *An Inquiry into the Good*, the 'unity of consciousness'. Within consciousness (or 'at its base [*sono kontei ni*]', a recurring phrase of Nishida's) is a 'unifying power ... [that] never exists apart from the content of consciousness; in fact, the content of consciousness is established by this power.'[217] It is a manifestation of what Nishida also refers to as 'a certain unifying something',[218] a woefully vague phrase but one that he intends as a definition of reality: 'true' reality is a single, dynamic, self-realizing *activity*, a unifying activity that brings together all the seemingly multiple realities of the world. And lest we miss the significance of his definition, Nishida repeats over and over again that the activity of reality encompasses the activity of human consciousness: 'the unifying power at the base of our thinking and willing,' he says, 'and the unifying power at the base of the phenomena of the universe are one and the same.'[219] Or again, 'A world that we can know and understand at all must be [a world] established by a unifying power identical to that of our consciousness.'[220]

The unity of consciousness has a shorter name: 'the self [*jiko*]'. Not 'self' in the usual sense of self-awareness, consciousness of our individual identity as an 'I-ness' or 'me-ness', because that derives from object logic and implies a false duality — an 'I/me' as opposed to a 'not-I/me'. 'True' self, like 'true' reality, denies duality: self is true only when we 'forget [*wasureru*]' or 'get rid of [*suteru*]' the 'I/me' self. Confusingly, Nishida has another name for the unity of consciousness: 'personality [*jinkaku*]'. Needless to say, he does not mean this term either in an everyday sense, how we behave, our characteristics, likes and dislikes, and so on. Although Nishida recognizes that 'an assortment of highly subjective hopes' [221] may express individual personality to some degree, essentially he uses 'personality' as a synonym for 'true self' — rid of all 'subjective hopes', and rid of all dualities — within the unity of consciousness:

> The true unity of consciousness is a pure and simple activity that arises naturally, unhindered by our usual sense of self [literally, 'our us-ness']; it is the original state of independent, self-sufficient consciousness, with no distinction between knowledge, feeling, and volition, and no separation of subject and object. This is when our true personality expresses itself in its entirety.[222]

That phrase 'the original state [*honrai no jōtai*]' is strangely evocative. Is it my imagination, or is Nishida here suggesting a variety of Fall — a Fall not of man but of consciousness? [223] The world we are familiar with is a world of distinctions, between things we know as opposed to things we feel or will; and between the supposed objects that we know and our knowing selves, the supposed subjects. And yet, Nishida is saying, despite all appearances to the contrary, all our commonsense assumptions about the world, these distinctions are not the facts of true reality. They are the perceived facts of a reality distorted by object logic, facts 'adulterated' by dualistic (or, at the risk of overstretching the analogy, fallen) thinking. True reality can only be found in the original state or true unity of consciousness, which is *prior to* all distinctions and identical to the unifying power behind the phenomena of the universe.

## The logic of *basho*

Whereas in *An Inquiry into the Good* reality had been located, so to speak, in pure experience, from the date (1917) of Nishida's second work [224] onwards the location is on the move. Initially it shifts to a 'self-awareness [*jikaku*]' that unifies 'reflection [*hansei*]' (which gives rise to object logic and other forms of dualism) with 'intuition [*chokkan*]' (in which the conscious human mind is directly and immediately aware of reality). Ten years later, intuition has taken on an even deeper significance and is now known as 'seeing without a seer'. The passage in which this startling phrase appears is quite a mouthful, but worth quoting in full because it indicates the growing importance of Nishida's notion of reality — identified with intuition — as a *self-determining and self-reflecting nothingness*:

> [What I mean by intuition is] seeing everything that is and everything that is at work as reflections of what mirrors itself within itself by itself becoming nothing. I want to conceive of a seeing without a seer at the base of all things.[225]

Mysterious as it may seem, 'seeing without a seer' is very closely related to the concept that is acknowledged, in Japan and more widely, to be Nishida's single most original contribution to philosophy: '*basho*', or 'the logic of *basho*'.[226]

Nishida was first impelled to re-examine and develop his thinking by his dissatisfaction with pure experience as he had originally expressed it. There was no change in the task he had set himself — to conceive of the facts of reality, true reality, directly and unhindered by dualistic thinking of any sort. However, as he noted somewhat defensively in his preface to the 1936 reprint of *An Inquiry into the Good*, 'the standpoint of this book is that of consciousness, which might be thought of as a kind of psychologism.'[227] In other words, Nishida was concerned that pure experience might be mistaken for the position that a theory of psychology is capable of explaining non-psychological facts or laws in epistemology, metaphysics, logic, and even mathematics. (Was he troubled, too, that pure experience might be perceived as a form of solipsism, the position that the contents of our own consciousness are all we can know, and may be the only reality?) [228] Nishida's other concern was that his philosophy was vulnerable to the charge of mysticism. To some extent, Nishida himself invites the charge through his seemingly casual use of the word 'mystical [*shinpiteki*]', as in phrases such as 'mystical intuition',[229] 'world of mystical intuition',[230] or 'a certain mystical something'.[231] Is this mere looseness of language, or ambivalence as to the validity of a mystical element within philosophy? Whatever the answer, by the time of Nishida's last essay he is at pains to say that his philosophy is *not* mystical, and that those who find it mystical are seeing it from the erroneous viewpoint of object logic.[232] It is significant that Nishida now makes a direct connection with Zen Buddhism, suggesting that misconceptions of Zen can be traced to the same cause: 'all misunderstandings of Zen [likewise] stem from thinking based on object logic.'[233] In truth, Nishida means to say, the mysticism charge falls away once it is grasped that his philosophy, like Zen, does not see the world — and cannot be seen itself — in terms of subject and object, or any other dualism. Just how this relates to the logic of *basho* (and why the connection with Zen is so significant) will become clearer before too long.

To defend his philosophy against accusations of psychologism or mysticism, Nishida came to feel, he would have to give it a 'logical foundation'.[234] He began to do so from the time when, as he later put it, he 'turned to the idea of *basho*'.[235] At first sight it looks as if Nishida is out to repudiate all his earlier philosophical standpoints, for there is nothing obviously logical about 'seeing without a seer', for example. On the contrary, Nishida himself has characterized it as a form of intuition — akin, it could be objected, to the mysticism that he is so keen to avoid. Nonetheless, there is a continuity here even with 'seeing without a seer'. The explanation, you will not be surprised to hear, is that Nishida has his own definition of 'logical'.

First a digression to say something about the etymology of the term '*basho*'. It is a perfectly ordinary word in modern Japanese, meaning 'place' or 'location.' However, Nishida invests it with complex nuances that are not easily captured. In English translations of his works, *basho* or aspects of it are sometimes rendered as 'place', but also as 'field', 'realm', the noun 'universal', the Latin word '*locus*', and the Greek word '*topos*'.²³⁶ Perhaps you see why I think it best to stick with *basho* and try to explain the nuances as we go along.

According to Nishida, the precursor to *basho* is to be found in Plato's *Timaeus*. It is the 'receptacle [*hupodochē*]' or 'space [*chōra*]' in which copies of the eternal realities, Plato's Forms, are made by a process of stamping or impressing, thus giving rise to the world of becoming and change.²³⁷ The receptacle is eternal, indestructible and endlessly plastic, never permanently altered itself by the impressions it imparts to the copies. It is this *featurelessness* of Plato's receptacle that seems to have caught Nishida's imagination, although, as we are about to find out, *basho* is featureless in a fundamentally ('logically') different sense. Nishida acknowledges his debt to Plato in an essay, published in 1926, simply entitled '*Basho*'. On the first page of the same essay is a sentence that translates as 'What is [i.e., exists] must be placed in something.'²³⁸ The 'something' now is *basho*, but merely giving it a name does not get us much further. More revealing is the fact that Nishida's expression for 'must be placed' is a modified form of '*oite aru*', the term he used (page 68 above) in his musical analogy. Far from combining to make a melody, he said, individual notes are already there — 'placed'— *within* it. Grasping that conception is the clue to understanding what Nishida means by the logic of *basho*. Or to put it another way, we cannot hope to understand the logic of *basho* without first grasping Nishida's notion that the part or particular (the note) is determined by and encompassed within the whole or universal (the melody).

While the musical analogy is still helpful, it would be better if we had something we could visualize. We're in luck because at times Nishida appears to use 'placed' with a more literal meaning, which allows him to conjure up this strikingly visual image:

> the infinite overlapping of universals upon universals, *bashos* upon *bashos*, an unbounded placement of circles within circles.²³⁹

In Plato's creation story, the realities of the world we know are imperfect copies of a perfect model, produced almost mechanically like shapes impressed into wax by a seal. There is nothing remotely mechanical about Nishida's concept of *basho*. *Basho* is not a receptacle, not even a space, but a cascade of reality that flows outwards, round and back, overlapping, encompassing, *enveloping* itself as it does so. Except that 'enveloping' is not quite Nishida's expression. His expression is '*hōronriteki*', which trips off the tongue in English as 'envelopingly logical'.[240] We have arrived at Nishida's definition of the logic of *basho*: *basho* is 'logical' in the sense that it envelops, and determines, what is within it. It is an 'enveloping logic'.[241]

## *Bashos* within *bashos*

For Nishida, there is only one reality, one whole or universal. We would expect, therefore, only one *basho*. Yet he speaks of more than one ('*bashos* upon *bashos*'), just as he seems to speak of more than one whole ('universals upon universals'). Nishida mystifies us and hints obliquely at what he means in the same short passage. Re-adjust your imagination, if you will, and focus on his image of concentric circles. Those circles within circles preserve the self-enveloping structure of *basho*, its wholeness, while permitting qualitative differences or gradations within it. The gradations themselves signify *degrees of reality* measured by inclusiveness (or *envelopingness*), ranging from least inclusive and least real to most inclusive and most real or, the term we are more familiar with, true reality. Crucially, whereas we might think of the innermost of the concentric circles as a solid and tightly bounded centre of reality, Nishida is adamant that the exact opposite is true. It is the outermost, all-encompassing and *un*bounded circle that is most real.[242]

*Basho*, then, can be one and more than one. Opinions differ as to how to distinguish their gradations (and what to call them),[243] but broadly speaking there are three major *bashos*.

The first, and *least* real, is 'the *basho* of beings [*u no basho*]' — of the individual beings (i.e., all things that exist, animate and inanimate) in the world and, by extension, the entire physical universe. On the one hand, it is downright peculiar to characterize the universe as least real, for by definition what can be more inclusive than the universe? On the other hand, there is a precedent of sorts for doing so: Plato's world of change, the visible, tangible world that comes to be and ceases to be, never 'fully real' because it is only a 'likeness' or 'moving shadow' of the eternal Forms.[244] The resemblance to Plato's world of change ends there, however, because Nishida's complaint against the world is not that it is a poor copy of a perfect model, but that it is not really made of what it appears to be made of. And though it is true that Nishida develops his own thinking partly through the 'mediation' (his word)[245] of Greek philosophy, it is to Aristotle rather than Plato he turns here, in particular Aristotle's notion of '*ousia*', the 'substance' or 'essence' that gives individual beings their identity. The beings that we suppose make up the world are what in everyday speech we would call objects. Yet in Aristotle's more precise sense, a logical and *grammatical* sense, each is not an object but a 'subject [*hupokeimenon*]', of which 'other things are said, but which is not itself ... said of any other thing'.[246] In other words, there is something in the subject, its *ousia*, that can never become the predicate of anything else. For example, we may say that a rose has the attribute of being red or white, scented or unscented, thorny or smooth, but we cannot say that redness, scentedness or thorniness have the attribute of being a rose. The problem, as Nishida sees it, is that while Aristotle professes to believe that individuals (substances, subjects, beings, things, or whatever we want to call them) are the basic units of reality, he makes them unknowable *in themselves* because all we do know are their attributes. For Nishida, a meaningful account of reality has to include our knowledge of reality and that, in turn, has to include our knowledge of individual beings in themselves. His solution to the problem is to invert Aristotle: to have an identity, says Nishida, Aristotle's subject must be identified with its attributes. To define or determine itself as the subject, that is, it must become its own predicate as well.

Inverting Aristotle sets Nishida's enveloping logic in motion. Everything moves outwards, not inwards, and the *basho* of beings is itself enveloped by another *basho*. Nishida has several names for this second *basho*, including 'the predicate-plane',[247] but for the sake of simplicity I shall refer to it as 'the *basho* of consciousness'. Whereas previously Nishida has spoken of consciousness (i.e., the self or personality) as a unity and unifying power, he now describes consciousness in terms of a 'field [*ishiki no ba*]' in which individual beings — the (grammatical) subjects that have become predicates — appear as the stuff of judgements and knowledge. By enveloping the individual beings, the *basho* of consciousness makes room for them — or, to borrow Heidegger's term, creates a 'clearing' in which they can exist and consciousness can judge and know them.[248] Moreover, consciousness has made a predicate of itself and although conventionally we might say 'I am conscious of X' (where 'X' is something I judge or know), Nishida would have us say 'X is what I am conscious of'. Why? Because for him the unity of the 'I' is not that of a (grammatical) subject but a 'predicating unity.... a circle rather than a point, a *basho* rather than a thing.'[249]

One of the other names for the *basho* of consciousness is 'the *basho* of oppositional nothing'.[250] To understand the nuance of 'oppositional', we have to start by reminding ourselves that Nishida's *basho* concept has a centrifugal dynamic. As they become ever more inclusive, the concentric circles spiral away from the centre and towards the outermost, unbounded *basho* — unbounded because it is an absolute nothingness and therefore cannot conceivably be enveloped by anything else. Although the *basho* of consciousness, too, is heading towards the outermost nothingness and provides an 'entrance [*iriguchi* or *kadoguchi*]' to it, the *basho* of consciousness is not, or not fully, a nothing itself because it is only a nothing in relation to the individual beings that it envelops, and 'opposes' in the sense of *not being* them.[251] A nothing in this relative sense alone, the *basho* of consciousness remains a quasi-being that is not properly self-determining: acts of consciousness within it are determined by the will, which in turn is guided by values or ideals that are reflections of the third *basho*.[252]

We have come to the most fundamental — and most real — *basho* of all, 'the *basho* of true nothing [*shin no mu no basho*]'. It is, so to speak, where the buck of determining stops, and stops in two ways. Firstly, the *basho* of true nothing is the field (or ultimate predicate-plane) in which *all* predicates, including those that determine individual beings within the *basho* of consciousness, are themselves finally enveloped. Secondly, while the *basho* of true nothing is not and cannot be determined by anything else, it determines itself — the buck stops — in the sense that it 'sees' itself reflected in all the determining acts that take place in the *basho* of beings and the *basho* of consciousness. This is recognizably intuition and 'seeing without a seer' (my page 72) again in a different guise. Sure enough, in the closing pages of his essay '*Basho*' Nishida speaks of '[t]rue intuition ... immediately placed in [the *basho* of true nothing] by breaking through ... the *basho* of consciousness', and refers to the *basho* of true nothing as 'the *basho* of intuition'.[253] The *basho* of true nothing is an intuition that, *within itself*, 'sees' itself mirrored in reality as a whole (or mirrored in 'everything that is and everything that is at work', as Nishida put it).[254] But as a seeing without a seer, the *basho* of true nothing is a nothing, an absolute nothing. By the same token, absolutely nothing can be said of it, not even that it is more featureless than Plato's *chōra*, for example, because that would be to try and make true nothingness into a something which has (or lacks) features.

Perhaps there is some merit in my own image (page 75) of the cascade of reality welling up from and returning to itself. For whereas concentric circles convey the centrifugal dynamic of the logic of *basho*, a self-recycling cascade conveys that the dynamic is also centripetal, or rather, circular.[255] Just as the cascade, on reaching the furthest point of its outward flow, would turn back on itself, so intuition, the outermost of the *bashos*, turns back and *inwards* to 'see' itself mirrored in the whole of reality. It is important to grasp, however, that while intuition 'sees' itself reflected in the beings and everything else it envelops, it does not cause or 'possess [*motsu*]' them, and nor does it act or 'work [*hataraku*]' upon them. It lets them be. '[W]hen beings are placed in true nothing,' Nishida writes,

> we have no choice but to say that the latter mirrors the former. To mirror means to let the thing stand as it is [*sono mama ni*], to receive it as it is, without distorting its form. That which mirrors allows things to stand within itself, but does not work upon them.[256]

Yet another way of expressing it is to say that the *basho* of true nothing is able to let things be because, in mirroring them, it negates itself to allow them to 'stand' as they are.[257] And its ability to do this is so limitless that it encompasses, *not resolves or reconciles*, all apparent opposites and contradictions:

> The *basho* of true nothing must be that which transcends the opposition of being and nothing in every sense and enables them to be established within itself.[258]

For 'opposition', of course, we could also read duality, and for the duality of being and nothing read part and whole, seer and seen, knowledge and knower, creation and creator, individual self and universe — any and every conceivable duality, not resolved or reconciled but enveloped in the greater unity of an unbounded, undeterminable, and self-reflecting nothingness.

I have chosen to focus primarily on the period from the publication of *An Inquiry into the Good* in 1911 to the '*Basho*' essay of 1926. Although Nishida's language changed and changed again within the span of those years, what I have tried to demonstrate in his thought is continuity rather than change. His abiding concern is remarkably constant. Pure experience, intuition, 'seeing without a seer', 'the unifying power at the base of the phenomena of the universe', and now the *basho* of true nothing — all these terms, and all the variants of them, keep taking us back to the first line of Nishida's first book:

To experience means to know facts just as they are....

Or as I have taken the liberty of rewording it:

To experience means to know reality just as it is....

Very simply (or very complicatedly, depending how you look at it), Nishida's whole philosophical project begins and ends with trying to know the nature of true reality. While this sounds like a recipe for abstract speculation, and much of Nishida's prose is indeed extremely abstract, the project itself is anything but abstract in its intent. On the contrary, the underlying restlessness of Nishida's writing suggests that something very insistent, bordering on compulsive, is driving him. The reason why he is so driven, I think, is that his is a moral as well as a philosophical project.

## The good

In Nishida — and this, too, surely reflects his preoccupation with unity in all things — there are no firm dividing lines between philosophy, religion, and morality. In his preface to the original (1911) edition of *An Inquiry into the Good*, Nishida had explicitly described religion as 'the consummation of philosophy',[259] and it seems he never wavered from that conviction. In 1928, the year after his realignment towards the *basho* concept, he described 'seeing without a seer' as a 'religious ideal [*shūkyōteki risō*]',[260] while the title of his last essay, 'The Logic of *Basho* and the Religious Worldview', speaks for itself. But it is not at all clear what he means by 'religious'. Albeit it is correct as the dictionary equivalent of '*shūkyōteki*', one suggestion is that 'existential' might be a better translation for the word as Nishida uses it.[261] In any case, we can be sure that it has nothing to do with a personal God standing outside the universe and controlling it, which Nishida brusquely dismisses in *An Inquiry into the Good* as an 'extremely infantile' notion.[262] One pointer to what Nishida does mean is to be found later in the same book: 'True religion seeks the transformation of the self and the reformation of life.'[263] In Nishida's early terminology, remember (pages 70-71 above), the self — 'true' self, not our shallow, everyday sense of self — is synonymous with personality and the unifying power of consciousness, which is itself identical with the unifying power behind all apparent realities. Hence the 'transformation' Nishida speaks of involves not only our relation with the reality of the self, but our relation with the reality of the whole universe.

If Nishida's identification of true self with true reality is inseparable from his conception of the religious, it is likewise inseparable from his conception of morality — which is to say, once more in the language of *An Inquiry into the Good*, from his conception of 'the good'. As the philosopher and Nishida scholar Abe Masao puts it, for Nishida the good is 'understood on the basis of reality'.[264] Or as Nishida himself puts it,

> to unite with the true reality of the self is the highest good. The laws of morality thus come to be included in the laws of reality, and we are able to explain the good in terms of the true nature of the reality called the self.[265]

The good — or 'good conduct'— does not consist in an abstract morality laid down by institutional religion, say, or the supposed demands of Reason with a capital 'R'.[266] Nor in 'egoistic hedonism' (i.e., the pursuit of individual pleasure, which Nishida regards as Epicurus's 'fundamental creed') or 'universalistic hedonism' (the pursuit of social or public pleasure, which he associates with Jeremy Bentham and the utilitarians).[267] The good consists in 'the development and completion — the self-realization — of the self'[268] or, as Nishida rephrases it yet again, in the expression of our 'innate nature':

> For a human to display his or her innate nature — just as a bamboo ... or a pine tree fully displays its nature — is our good.[269]

What follows is my personal interpretation and Nishida scholars may or may not agree with it, but here goes. The point of the bamboo and pine simile, which probably alludes to a saying attributed to the poet Bashō,[270] is that 'innate nature' cannot be understood in terms of 'human nature', that is, in terms of some essence or propensity believed to be common to all human beings. Our innate nature is not to be sinful, as a Christian theologian would say; or to feel driven to produce our own means of subsistence, as Karl Marx would say; or to be prone to aggression, as Konrad Lorenz would say. Even when, to a degree at least, some such characterizations of human nature happen to be true, they are orders of magnitude removed from what Nishida means by innate nature. What he means is *suchness*, the condition (for want of a better word) in which we are just as we are, and not otherwise. It is not the same as personality in the conventional sense of what we are like individually (although Nishida acknowledges it includes that). It is something much more profound and ultimately metaphysical. Of all the possibilities that have emerged and might ever emerge from the nothingness of *śūnyatā*, it is *our* possibility, our suchness, that has come into being: each of us is inextricably part of *śūnyatā* and yet distinctly, intensely, uniquely differentiated from it. Moreover, contrary to how it may sound, we do not have to be in some exalted, transcendental or mystical state to be just as we are and not otherwise. If only we knew it, our suchness is evident in the very ordinariness of our daily lives — the ordinariness of '*byōjōtei*', as Nishida calls it.²⁷¹ To illustrate *byōjōtei*, Nishida quotes the ninth-century Chan master Linji (known as Rinzai in Japan), who taught that Buddhism requires no special effort: 'You have only to be ordinary,' Linji is said to have told his followers, 'defaecating, urinating, dressing, eating, and lying down to rest.'²⁷² The profound and the metaphysical, then, are one with the banal and the earthy.

To express our inner nature is to know the self. To know the self is to know reality. To know reality is to know that

> the self and the universe share the same foundation; or rather, they are the same thing.... The truth, good, and beauty [*shin-zen-bi*] of reality must be the truth, good, and beauty of the self.²⁷³

That the self and the universe are one and the same reality is, of course, a Buddhist tenet, as Nishida himself observes just before the lines I have quoted. And although elsewhere (with the notable exception of his final essay in 1945) Nishida appears to have been engaged primarily with Western thought, the core of his own philosophy-cum-religion-cum-morality is unmistakably Buddhist, and Zen Buddhist in particular.²⁷⁴ Which brings me back where I began. Nishida might have done better, I first suggested, to heed the warning — made by Huineng, the illiterate Chan patriarch — that the written word is of little use as a means to knowing true reality. On reflection, I end by suggesting instead that, however convoluted his language and imperfect his achievement, we have to admire Nishida for attempting to narrow the gulf between the expressible and the inexpressible, words and *śūnyatā*. To admire him, I mean, for the spirit of his lifelong project to teach Huineng to read and write.

# 7. Conclusions: time we knew the world aright

I began this whole essay with Nagel's pronouncement that 'everything we believe ... has to be based ultimately on common sense, and on what is plainly undeniable.'[275] That, I said, is an assumption, not a self-evident truth. So far, so good. You may object, however, that it is unwarranted to go further, as I do, and assert that not only is Nagel's trust in common sense unfounded, so too is his trust in the intelligibility of the world itself. I agree it is an extreme assertion, because it means that we have no reason or right to expect the world to make sense to us. I do not agree that it is unwarranted, because it is founded on more than a century's worth of robust scientific evidence telling us that fundamentally — at the scale of the very small (quantum) as well as the very large (cosmological) — the world is *un*intelligible. Plainly and undeniably unintelligible. I have explored the evidence at length (pages 9-16) and come back to it again below. But let's stay with your objection a while longer. Surely there must be some way of holding onto our belief in the intelligibility of the world?

If we choose to set aside any requirement for hard evidence, we are in the realms of faith or metaphysics. Nagel himself goes in the direction of faith — not faith in a God but a boundless and, dare I say it, dogmatic faith in the intelligibility of the world. As a contribution to what he calls, remember, 'the larger project of making sense of the world', Nagel sets out in search of alternative explanations that 'make mind, meaning, and value as fundamental as matter and space-time in an account of what there is.'[276] He is reconciled to the possibility that the fullest explanation might have to include elements of 'natural teleology', which, he speculates,

> would mean that the universe is rationally governed ... not only through the universal quantitative laws of physics ... but also through principles which imply that things happen because they are on a path that leads toward certain outcomes — notably, the existence of living, and ultimately of conscious, organisms.'[277]

In turn, this explanation of the world would have two related consequences. Firstly, we would come to realize that '[e]ach of our lives is a part of the lengthy process of the universe gradually waking up and becoming aware of itself.'[278] Secondly, and even more momentously:

> Such an explanation would complete the pursuit of intelligibility by showing how the natural order is disposed to generate beings capable of comprehending it.[279]

Nagel's conclusion, it appears, is that the world exists the way it does so that we are here to comprehend it, and so that we can comprehend it. Or as he expresses it himself:

> It seems to me that one cannot really understand the scientific world view unless one assumes that the intelligibility of the world ... is itself part of the deepest explanation of why things are as they are.[280]

If Nagel wants to 'complete the pursuit of intelligibility', though, why stop there? Why not reconcile himself to the possibility that the universe exists the way it does (and perhaps exists at all) because *that is how we ourselves have made it*.

There is a precedent for such a mind-boggling idea and, a nicely ironic twist, it is consistent with 'the scientific world view' that Nagel professes to respect. In 1979 the physicist John Wheeler was one of the speakers at a symposium held in Princeton to celebrate Einstein's centenary.[281] Wheeler suggested that the universe is as it is and has 'tangible "reality"' because, long after the Big Bang, beings with complex consciousness (ourselves, the only conscious beings we know of that would fit the bill) evolved to the level where they could make observations of the universe and, by those very observations and in 'a strange inversion of the normal order of time', bring it into existence.[282] Wheeler was reasoning from a quantum thought-experiment of his own, the so-called delayed-choice experiment, and from there to a concept he called 'the participatory universe'.

At this point it helps to recall (my page 33) one of the basic principles of quantum mechanics: because the act of observing a quantum system (i.e., any system that involves quantum objects, such as photons or electrons) changes the system, it follows that the observational apparatus and the experimenter (who has chosen what to observe and set up the apparatus accordingly) are inescapably part of the system, observer-participants. Now the universe is very big but ultimately it is made of things so small that they are themselves quantum objects, and hence subject to observer-participancy. According to Wheeler, there is a 'mechanism' (his word) by which observing the behaviour of quantum objects would have consequences at the cosmological scale: his delayed-choice experiment, which is a modified version of an experiment routinely carried out in the laboratory. The standard laboratory version consists of two screens, the first perforated with two slits and the second incorporating a detecting device, placed one in front of the other such that when photons are fired through the slits they leave a characteristic pattern on the detector screen. Even in the standard version the experimenter determines, in effect, whether the photons — including *individual* photons — will behave as though, in a particle-like manner, they have entered the apparatus through one slit or, in a wave-like manner, through both slits at once.[283] Wheeler's modification of the standard apparatus does not change any of that, except in one respect: it allows the experimenter to delay his or her choice of which experiment ('through one slit' or 'through both') until *after* the photons have entered the apparatus but before they reach the detecting screen. Depending on the choice, the photons will still behave as they did before, the difference being that now they do so under conditions that effectively have been determined after the event.[284] This is the 'mechanism', applied at the scale of the whole universe, that led Wheeler to suggest in 1979:

> In the delayed-choice experiment we, by a decision in the here and now, have an irretrievable influence on what we will want to say about the past.... This [strange inversion of time] reminds us more explicitly than ever that 'The past has no existence except as it is recorded in the present'.... [F]or the building of all law, 'reality' and substance ... what choice do we have but to say that in some way, yet to be discovered, they all must be built upon the statistics of billions upon billions of ... acts of observer-participancy?[285]

It is difficult to gauge Wheeler's intent from the text of his presentation alone. How literally or seriously, at the time and later, did he himself believe what he said? I don't know the answer, but I would like to think that he speculated on a participatory universe as a way of asking how far we are willing to go to preserve our belief — or faith — in the intelligibility of the world. To say that human beings have given the universe its 'tangible "reality"' is, it seems to me, akin to saying that the universe did not exist until the day in 1964 when Penzias and Wilson detected the cosmic background radiation, and do we believe that?

I rest my case against Nagel's kind of faith. While we are considering alternatives to evidence-based understandings of the world, however, and in the interests of balance, maybe we should try going in the direction of metaphysics. All right, we will. Some time ago (page 40), apropos Sartre's misinterpretation of Heidegger's *Being and Time*, I suggested that not only did Sartre turn Dasein into consciousness (the 'for-itself [*le pour-soi*]' in Sartre's terms), he provided consciousness with a creation story of its own. I look again at his creation story now because it offers a metaphysical explanation of consciousness and, by extension, of our compulsion to try and make sense of ourselves and our world.

The primeval matter of Sartre's creation story is the 'in-itself [*l'en-soi*]', the inanimate, non-conscious content of the world. Unlike the for-itself, which is always 'at a distance from itself',[286] the in-itself is perfectly identified with itself, 'glued to itself [*s'est empâté de soi-même*]'.[287] Inert and 'solid [*massif*]' in the sense that nothing about it is hidden or could be other than it is, the in-itself 'exhausts itself in being'; worst of all, it is contingent, *just there* without any reason for its being or connection with other things, meaningless, '*de trop* [superfluous] for eternity'.[288] The crux of Sartre's creation story is that, anthropomorphically speaking, the in-itself has a great desire: it wants to 'found itself [*se fonder*]' and 'remove contingency from its being'[289] — in short, to justify its own existence. And to do so it 'recovers itself [*il se reprend lui-même*] by degenerating into'[290] a for-itself, consciousness, because consciousness is uniquely capable of being aware of things and providing reasons for their existence. 'It is only by making itself for-itself,' Sartre says, that the in-itself 'can aspire to be the cause of itself.'[291] In the same passage he speaks of 'a rupture in the identity-of-being' of the in-itself, but elsewhere his imagery waxes biblical and he refers to 'this fall of the in-itself toward the self, the fall by which the for-itself is constituted'.[292] Here Sartre's consciousness-creation story not just invites comparison with the biblical Fall, it *rewrites* the Fall. In the beginning was inanimate matter. But inanimate matter, in an urge to justify its own existence, ruptured the perfect identity of itself with itself and fell into consciousness. Once conscious, and driven by the same urge, matter sought to justify — make sense of — the world in which it found itself (only to discover, in the sequel to Sartre's story, that the world itself is contingent, *de trop*, and absurd).

Up until the point where Sartre rewrites the Fall, he and Nagel seem to be saying much the same thing, namely that there is a reason why we have conscious awareness: according to Sartre, inanimate matter's horror at its own contingency; according to Nagel, a reason relating to 'the lengthy process of the universe gradually waking up and becoming aware of itself'. The big difference between them, however, is that whereas for Nagel we have consciousness because the universe exists *for the purpose of* 'generating' beings capable of comprehending it, for Sartre we have consciousness precisely because everything is contingent and the universe has *no* purpose at all.

Doubtless Heidegger would have dismissed Sartre's consciousness-creation story as yet more worthless metaphysics that fails to address the real issue (for Heidegger, the issue of Being). Personally speaking, I find Sartre's creation story oddly appealing — perhaps because there is an almost mythological quality to it, certainly because it is free of teleological implications. At the end of the day, though, and whether we call it metaphysical, mythological or anything else, Sartre's story is no substitute for the hard, scientific, evidence-based understanding of the world with which we began, and which we must confront if we are to recover the 'natural' knowledge we need. The knowledge, that is, of the world itself, uncreated, without purpose, and ultimately unintelligible.

## The world is as it is, but it didn't have to be

It was God who made the world, Genesis tells us, albeit it does not tell us why. Plato is more forthcoming: the world exists because the creator was so lacking in envy that he wanted all things to be as like himself as possible. In a sense, and counter-intuitively, the fact that Genesis does not specify a motive for the creation makes the biblical account closer to the truth. For we now know that the true creation story goes like this. In the beginning was nothing. Then, 13.8 billion years ago, there was something. Yet there was no creator, and no motive, design or purpose to explain why the world came into being the way it did, or came into being at all.

To recap, currently there are three main pillars to the observational evidence that our universe burst into existence at an arbitrary instant in the distant past. First there is the fact that the universe is expanding (at a velocity of the order of 70 kilometres per second, and accelerating): reversing the expansion tells us that 13.8 billion years ago everything that now comprises the universe would have been compressed into a single point, extremely hot and infinitesimally small. Secondly, the distribution throughout the observable universe of the lightest elements, hydrogen, deuterium, helium and lithium, agrees well with statistical predictions of which elements would (and which would not) have resulted from nuclei formed in the Big Bang nucleosynthesis. Third is the cosmic microwave background radiation, the uncanny afterglow of the Big Bang itself — now down to 3 degrees above absolute zero, the temperature predicted by theory. These three pillars of evidence are so fundamental that they should be as much part of general knowledge as the names of the planets, for example, or major rivers and capital cities on earth. I doubt my recap will help towards that end, but something shorter and catchier may have more chance of success:

> Once there was nothing to see,
> Then something the size of a pea;
> It started quite lightly
> But grew expeditely,
> All the while cooling to 3.

I appreciate that you may still prefer a figurative creation story to the brute facts of physics and cosmology. If so, and even if you find Genesis or the *Timaeus* wanting in this respect, all is not lost: Lucretius's *De rerum natura* has a good claim to being the most complete and most truthful of such creation stories. Complete, I mean, in the sense that it is fully worked out; truthful in the sense that it acknowledges — celebrates, indeed — the *randomness* of the process by which the universe came to exist. As we saw earlier (page 22), Lucretius stresses the randomness of it all because, like Epicurus, he wants to escape the rigid determinism of the original Atomist physics and philosophy. The randomness is significant in its own right, however, because it belies the notion that the process of creation had anything deliberate or inevitable about it. Quite the reverse: 'our world,' Lucretius insists, 'has been made by nature through the spontaneous and casual collision and the multifarious, accidental, random and purposeless congregation and coalescence of atoms.'[293] Spontaneous, casual, accidental, random, purposeless — not words we would ever use of a universe created by design, least of all the design of a God or gods.

An unexpected bonus of Lucretius's creation story is that it both rings true imaginatively and is broadly consistent with what we know of how the early universe actually developed, from the moments after the Big Bang (which does not, of course, feature in Lucretius's poem) to the formation of the first stars about 100 million years later. In the beginning, Lucretius says, was nothing but 'a newly congregated mass of atoms of every sort.'[294] It was these 'multitudinous atoms, swept along in multitudinous courses through infinite time ... [that came] together in every possible way and tested everything that could be formed by their combinations.'[295] Allowing for the fact that Lucretius could not have conceived of a nucleus or anything else smaller than an atom (indivisible by Atomist definition), he might almost be describing the early phases of chemical evolution. As for the later life of the universe, Lucretius tells us that although the atoms themselves are indestructible, all the varieties of matter they go on to create, including the earth and the stars, are perishable and one day will come to an end. That, too, we now know to be true.

While the poet-philosopher Lucretius may give us the best figurative account of how the world came about, to some extent Nishida's concept of the *basho* of true nothing (or 'the nothing' for short) also offers an alternative to the bare facts of physics and cosmology. Nishida does not tell a creation story as such because he is silent as to how the world came to exist in the first place. Nonetheless, his account of the world as it is poses an intriguing challenge to our conception of a creator, and belatedly resolves a conundrum that Western philosophy and theology puzzled over for more than two thousand years:

- **A creator God is a contradiction in terms.** The notion that God made heaven and earth necessarily implies the duality of a creator and creation. But the nothing is beyond all dualities — not only of creator and creation, even of nothingness and somethingness. Whatever else Nishida may understand by the concept of a God,[296] by the logic of his own concept of *basho* there can be no creator God because, unlike the nothing, such a God would not have 'completely shaken free from the character of being' and would therefore be tainted with 'somethingness'.[297] Notwithstanding Nishida's apparent lack of interest in the origins of the world, implicitly he rules out divine intervention as a way the world could have come into existence.

- **The relation between the one and the many resolved.** Especially under Plato's influence, the conundrum of the one and the many asks why the world is so full, varied and mutable when there might just as well be a single, uniform and eternal Idea of a world. According to Nishida, however, there is no contradiction or conflict between the particular (i.e., all the things or beings in the world, which are differentiated and many) and the universal (the nothing, which is universal and one in that it is undifferentiated).[298] By the very fact that it negates itself as it envelops the whole of reality, the nothing enables particular things or beings to exist as they are in all their distinctness and immediacy.[299] The nothing does not cause, 'possess' or 'work upon' them but sees itself mirrored in them, *a seeing without a seer*. For to mirror, as Nishida has told us, 'means to let the thing stand as it is, to receive it as it is, without distorting its form.'[300]

## The world is under no obligation to make sense to us

I have defined an intelligible world as one that has, or is assumed to have, some reason for being as it is, or for existing at all. What Lucretius, Bohr, Heidegger and Nishida have in common is that none of them seem to have assumed the world to be intelligible in this sense. On the contrary, and each in their own way, they all suggest to us that the supposed intelligibility of the world is an illusion that, at best, impedes a true understanding of the world and, at worst, prevents it altogether.

At first sight, Nishida appears to be the odd man out. We could be forgiven for taking his 'unifying power at the base of the phenomena of the universe', synonymous with the 'unifying power of consciousness', to imply a reason or purpose behind everything — which, if discoverable, would confirm that the universe is intelligible. In the same work, *An Inquiry into the Good*, Nishida writes:

> Will is the deepest unifying power of our consciousness and also the most profound expression of the unifying power of reality…. [N]ature is an expression of the will, and it is through our will that we can grasp the true significance of profound nature.[301]

And one consequence of our possessing the 'infinite unifying power of reality' is that 'we can search for the truth of the universe in learning'.[302] While this sounds like an invitation to study the world as evidence of purpose (God's purpose, Bacon and Milton would insist), it is nothing of the sort. For Nishida, as I understand, the notion that the universe has some purpose is inextricably linked with a meaningless notion of God. As he puts it, to 'infer an omniscient controller from the fact that the world is organized favourably according to a certain goal, one must prove that the myriad things in the universe are in fact created purposefully,' which, he observes with admirable understatement, 'is extremely difficult to do.'[303] The appearance is deceptive. The world, Nishida is saying, is as it is *in its suchness*, not according to any design or purpose. A suchness determined by design or purpose would, like a creator God, be a contradiction in terms.

If there is an odd man out, it is Heidegger. For him, too, the world is ultimately unintelligible but the world he means is the they-world (my phrase, not Heidegger's), the society and culture into which Dasein has been born, socialized, and 'dispersed'. Dasein is under the impression that it has an identity and meaningful grounding of its own because it perceives that it, Dasein, determines the possibilities into which it presses forward and how it understands itself. In fact, Heidegger insists, Dasein has no identity or grounding of its own because all its possibilities, together with its understanding of itself, are determined by publicness and the 'they'. Such is Dasein's — that is, our — big illusion and we are only forced to confront it, rarely if ever, in the mood of anxiety. For in anxiety, Heidegger tells us,

> we are 'in suspense'. More precisely, anxiety leaves us hanging because it induces the slipping away of beings as a whole.... In the altogether unsettling experience of this suspendedness where there is nothing to hold onto, pure Da-sein is all that is still there.[304]

Anxiety reveals that what we are suspended in (or 'held out into') is *the nothing*, which is to say, we are nothing but the possibilities and average understanding dictated by publicness and the 'they'. Yet publicness and the 'they' cannot be located or pinned down in any way:

> The 'they', which is nothing definite, and which all are, though not as the sum, prescribes the kind of Being of everydayness.... Publicness primarily controls every way in which the world and Dasein get interpreted, and it is always right.... It 'was' always the 'they' who did it, and yet it can be said that it has been 'no one'.[305]

We are being told that our sense of deciding for ourselves what we do with our lives, and our sense of individual identity, our 'I-ness', are illusions of the they-world in which we happen to find ourselves.

Worse, publicness and the 'they', which create and perpetuate these illusions, are unintelligible to us on two counts. Firstly, while publicness has the force of law (is 'always right', as Heidegger expresses it), it is based on no more than arbitrary custom and convention, has no particular reason for being as it is, and could well be different. Secondly, the 'they' is so elusive that it cannot be identified as a who or a what in this place, that place, or any place at all. Hence publicness and the 'they' are themselves a kind of nothing, present everywhere and nowhere, incapable of being explained in terms of a something somewhere that would make sense to us.

Lucretius is refreshingly direct about the unintelligibility of the world — and he means the whole universe, not just the human world. In 'the initial concentration of matter,' he tells us, 'the atoms did not post themselves purposefully in due order by an act of intelligence, nor did they stipulate what movements each should perform.'[306] So there. Nothing purposeful, no acts of intelligence, and no stipulations. The gods exist, Lucretius concedes, but even if they had a plan or purpose for the world (which they do not), they could never implement it because the atoms have an independent and eternal existence of their own. An intelligibility of sorts — a crude cause-and-effect sort — might have been salvageable had Lucretius reverted to the rigid determinism of the early Atomists (the 'inescapable and merciless necessity' propounded by Democritus: my page 22). But Lucretius does not revert: following Epicurus, he invokes the 'swerve' to preserve the randomness and unpredictability of the atoms' behaviour, and in so doing he leaves no prospect of orderliness or consistency. Where, we ask, is the scope for intelligibility in a world made of atoms that casually collide, combine, break apart and recombine in different ways as they whirl about in infinite space?

The single most persuasive proponent — I'm tempted to say the prophet — of the unintelligibility of the world is Bohr. Not only does he have the authority of speaking as one of the founding fathers of quantum physics, he is prepared to acknowledge the evidence of unintelligibility without equivocation and however much it may offend 'reasonable' assumptions about how the world works.[307] In Bohr's case the offending evidence was the behaviour of quantum objects, which led him to call for 'a final renunciation of the classical ideal of causality and a radical revision of our attitude towards the problem of physical reality'.[308] Bohr did not expect such evidence, or the philosophical implications of it, to be confined to the lecture hall or specialist journals. On the contrary, he was optimistic that in time complementarity (the principle that position and momentum, like waves and particles, must be described in terms of complementary properties: my page 33) would come to be generally accepted, and not just within scientific circles but in a broader educational and cultural sense. In response to a suggestion that science cannot offer the 'guidance and consolation' offered by religion, for example, Bohr is said to have declared — 'with intense animation' — that he foresaw the day when complementarity would be 'taught in the schools' and that, 'better than any religion, ... a sense of complementarity would afford people the guidance they needed.'[309] Likewise, in the last interview he ever gave (literally the day before he died in November 1962), Bohr spoke of his confidence that 'the complementary description' would become part of 'common knowledge'.[310] In the same interview, and clearly more in sorrow than in anger, Bohr came close to accusing Einstein and Planck (who, like Einstein, refused to accept the epistemological implications of quantum mechanics) of wanting to *unlearn* knowledge that they themselves had brought into the world. Planck, Bohr said, 'used the last forty years of his life, not to say fifty, to try to get his discovery [i.e., as I take Bohr to mean, the quantum of action or Planck's constant] out of the world', while Einstein 'did this wonderful thing and a very, very fine paper, that in 1917, but since that time he tried to get it [again as I understand, the concept of quantization, one of the cornerstones of quantum physics] out of the world.'[311]

Similar sentiments have been expressed by others, perhaps most vehemently by the French physicist and philosopher of science Bernard d'Espagnat. In his contribution to a major international symposium in 1972,[312] d'Espagnat had sharp things to say about the fact that separability (or locality, the assumption that there can be no influence between spatially separated quantum systems: page 32) continued to be taught even though 'we may safely say that *non-separability* is now one of the most certain general concepts in physics'.[313] Fifty years on, d'Espagnat's presentation retains much of its force. The focus of his complaint is what he calls the 'multitudinist' conception of nature, 'according to which the ultimate reality — all that really is — would essentially be constituted by an enormous number of elementary events and/or microscopic objects, each one of them being endowed with simple properties and being such that the interactions of them all — taken as local and causal — would, combined with chance, give rise to the complexity of appearances.'[314] It seems that d'Espagnat attributes responsibility for this 'false view of nature'[315] to Democritus and the Atomists, which is disconcerting given everything, mostly positive, that I have said about them. For d'Espagnat, however, the real villains of the piece are not the Atomists, who can hardly be blamed for knowing nothing of Planck's constant, but the physicists of his own time, who know a lot about Planck's constant and yet, for teaching purposes, go on behaving as if they did not. Pointedly addressing himself to his audience of eminent peers and colleagues, d'Espagnat concludes:

> On the question that [the 'multitudinist'] view of the world is but a crude model *tailored to specified purposes of ours*, an absolute silence is nearly always kept. This mutilation of the truth throws undue discredit on the concept of *unity*, and is therefore a really serious one, culturally I mean. Moreover, it is a falsehood recognized as such by, [I] guess, a vast majority of the participants in this symposium…. And falsehoods should not be spread.[316]

Complementarity and non-locality are so troubling because they directly contradict two commonsense notions — that it must be possible to tell where something is *and* how fast it is moving, and that it is absurd to suppose a particle can somehow 'know' what another particle is up to on the other side of the universe. Nevertheless, Bohr and d'Espagnat are saying, that is the nature of 'physical reality' at the quantum scale and, sooner or later, educationally and culturally, we have to reckon with it.

As far as the educational side of the reckoning is concerned, I have no qualifications to suggest how physics should be taught anywhere or at any level. I will make one observation, though. Sixty years after Bohr's death, his prediction that complementarity would be 'taught in the schools' is at last coming true. According to a relatively recent (2019) international study of physics curricula, complementarity or 'wave-particle duality' is included in the physics courses, mostly elective, at upper secondary level (typically for students aged 17-19) in 13 European counties, plus Australia and Canada.[317] Germany seems particularly enlightened. Aspects of quantum physics have probably been taught in secondary schools in some German federal states since the 1960s,[318] while in Bavaria complementarity is now included in the general (i.e., compulsory) physics course for all students aged 15-16. In the spirit of Bohr's optimism, let's hope that the Bavarian curriculum will encourage educators elsewhere to make complementarity (and, appropriately taught, other key concepts of quantum physics)[319] part of the general education — Bohr's 'common knowledge' — available to everyone.

On the cultural side of the reckoning, we have to start by recognizing that the world we think we know is nothing but a convenient illusion, a bubble of familiarity, that derives from our own sense of what is 'real' and 'reasonable' in everyday things and events. Yes, the moon is there whether we are looking at it or not. Yes, tables and chairs look like tables and chairs and do what tables and chairs are supposed to do. And yes, we can be sure that every other entity we encounter in the everyday world, from toothbrushes to jumbo jets, will appear and function as we expect. But fundamentally there is no reason — no classical, causal reason — why they should because, in common with everything else in the universe, the moon, tables and chairs, toothbrushes and jumbo jets ultimately consist of matter and energy subject to the uncertainty and unintelligibility of quantum phenomena. It is an illusion of scale, or rather, of the disparity between the macroscopic scale of the world we inhabit and the microscopic scale (defined by Planck's constant) at which quantum phenomena occur.

Comfortably trapped within the bubble of our illusion, it is very difficult even to recognize that it is an illusion, let alone try to break out of it. It is worth remembering, however, that we have confronted such an illusion before, and successfully freed ourselves from it. As late as the sixteenth century, we cheerfully placed the earth at the centre of the universe and had the sun revolve around the earth. It was an easy mistake to make, for when we see the sun rise in the east and set in the west, as we still say, our unthinking perception is that we are indeed witnessing the motion of the sun relative to the earth. It took the meticulous observations and calculations of Copernicus, Tycho Brahe, Kepler, Galileo and Newton to convince us that what we are actually witnessing is the motion of the earth relative to the sun. Yet the mistaken belief that the sun went round the earth did not affect how most people lived their lives, or how larger histories unfolded. Farmers got on with ploughing, sowing and harvesting, and clerics with compiling holy calendars, untroubled by their erroneous understanding of what the sun was up to. Meanwhile explorers, traders and invaders reached their destinations using the astrolabe and other navigational methods that implicitly assumed an earth-centred universe. That is the point of the analogy. The geocentric model of the universe *worked* for most practical purposes at the time, just as our familiar model of the everyday, macroscopic world works for most purposes today.[320] Conversely, just as its lack of practical consequence did not save the geocentric illusion from Copernican astronomy, so the fact that our intelligible-world illusion lacks obvious consequence will not, I think, save it from modern physics and cosmology. It is a question of when, not if, the bubble finally bursts.

It is unlikely, or so we hope, that trying to convince people of the unintelligibility of the world will meet the sort of ferocious resistance that Copernican astronomy once met (so ferocious that, as Bohr puts it, 'Bruno was absolutely killed, and Galilei [sic] was forced to recant').[321] The bad news is that in other respects the Copernican analogy is flawed and misleading. John Donne's 'And new philosophy calls all in doubt'[322] hints at the bewildering impact of the so-called Copernican Revolution in early seventeenth-century Europe, and we can well imagine that the new astronomy seemed not only shocking but downright unintelligible. What possible reason or purpose could God have had for creating a universe that did *not* locate the earth at its centre? It made no sense. But therein lies the flaw in the analogy. Although our ancestors may have found Copernican astronomy unintelligible, in reality it was not. As Galileo and Newton in particular went on to demonstrate, the heliocentric model makes very good sense according to the laws of gravity and planetary motion. The unintelligible turned out to be intelligible after all. Hence while the Copernican analogy is encouraging in that it shows we are capable of liberating ourselves from a very old collective illusion, ultimately it is not applicable to the Big Bang or quantum phenomena, which remain obstinately and irredeemably unintelligible.

**We are of the world, and of the nothing**

For Heidegger, to say that we are in the world is to describe our state of Being, not our whereabouts. Implicit in his terms 'Being-in-the-world' and 'Being-in', the distinction is clearer in 'Being-already-amidst-the-world' and 'Being-amidst'. We are not just in the world in a spatial sense, we find ourselves amidst it, *already* amidst it, and not of our own accord but thrown there.[323] This is the state of Being that Heidegger calls Da-sein ('Being-there', which '*is* in such a way as to be its "there"' and 'brings its "there" along with it')[324] and marks his irreversible break from the Cartesian fiction that we are always in a detached and self-contained relation of thinking-stuff subject to the extended-stuff objects of the world. While Heidegger acknowledges that there is a place (and in scientific theory, for example, a crucial place) for subject/object thinking, he argues that it is not our commonest or most characteristic relation to the world — not how we encounter the world 'all the time in our experience'.[325] Most commonly and most characteristically, we are inseparable from the world we inhabit. We are both in the world and *of* it.

For all these reasons, albeit with one caveat, I propose that 'Being-already-amidst-the-world' — the concept and even the cumbersome term itself — should be a key component of the 'natural' knowledge we need to recover. Or as Heidegger would say, the knowledge we need to recall.

The caveat relates to Heidegger's conception of the world. He defines the word 'world' in several different ways, most importantly as 'that "wherein"' Dasein 'lives' and encounters (or more correctly, is encountered by) beings and entities other than itself.[326] Although he describes it in unfamiliar terms, Heidegger's world otherwise seems more or less familiar. The beings of Dasein's encounters are other human beings (other Daseins), while the entities include inanimate things like doors and door handles, hammers and nails. All is not as it seems. The beings and entities constitute Dasein's world by providing the background against which Dasein's everyday activities appear meaningful, the background (or 'significance' in Heidegger's vocabulary) that 'makes up the structure of the world — the structure of that wherein Dasein as such already is'.[327] But the background of meaningfulness only constitutes a world in a sense akin to that of expressions such as 'the business world', 'the theatre world' or 'the world of politics', that is, as a closed system of activities, conventions, values, and so on, common to a particular community. In this case the community is the whole of humanity, yet it still does not make it a world in a sense that would work for a genuinely 'natural' knowledge. Take inanimate things, for instance. In the world as we experience it most characteristically, Heidegger tells us, we do not encounter door handles and hammers as such but as 'equipment' that may be 'ready-to-hand' or 'un-ready-to-hand' (my page 39) in relation to some 'for-the-sake-of-which' (page 47) that is consistent with the understanding of Being we share with the 'they'.[328] Even the liberation we might hope to achieve as resolute Dasein (pages 55-56) would not enable us to escape (or 'float above')[329] the they-world, only to escape the illusion that we have an identity or significance of our own within the they-world. Heidegger's great insight is that we are of the world as well as in it, but by his definition it remains an anthropocentric (or they-centric) world, which is far too narrow a conception of *what is* — everything that exists — and our relation with it.[330] We need a better, broader definition of 'of the world'.

A better definition has to start from the reality that we are made of the same material as the world: Lucretius's multitudinous atoms or, more strictly, the chemical elements synthesized within ancient stars and then scattered throughout the universe when those stars burnt out and died. We are, quite literally, made of star dust. While we may think of this as something we learnt in the twentieth century, in fact we could and would have known it for much longer if only we had registered the fundamental truth of the Atomist picture of the world celebrated by Lucretius in the first century BC. Although the Atomists were wrong about the indivisibility of the atom, they were absolutely correct in their understanding that we share our origins with all matter, animate and inanimate. In the beginning there was no duality of thinking stuff and extended stuff. There was just stuff.

At the quantum scale of matter and energy — and that really is something we only began to grasp in the twentieth century — we are of the world by virtue of being its observer-participants. I mean observer-participants not in Wheeler's speculative sense that we have given the whole universe its 'tangible "reality"' by observing it (page 84) but in the factual sense, definitively expressed by Bohr and routinely demonstrated in the laboratory, that 'any observation of atomic phenomena will involve an interaction with the agency of observation'.[331] The agency of observation includes, of course, the human beings who choose which quantum experiment to perform and, by the way they set up the apparatus, define the conditions under which the phenomena will appear. As the dividing line between the microscopic scale of quantum mechanics and the macroscopic scale of classical physics is clearly demarcated by Planck's constant, it seems safe to say that observer-participancy in Bohr's sense does not apply at the everyday, macroscopic level. It is possible, I imagine, that one day Planck's constant may prove to be less clear-cut a demarcation than we have thought.[332] In the meantime, and metaphorically at least, observer-participancy is a potent reminder of our state of being: both in the world and of it.

How are we doing? Our definition of 'of the world' now begins with the recognition that all things, including ourselves and the world itself, share a common origin. The definition extends to our everyday experience of the culture and society into which we happen to have been born. But it cannot end there if it is to be a meaningful part of our 'natural' knowledge for a secular world. It must go beyond Heidegger's anthropocentric they-world and comprehend the reality of *what is* directly, fully, and honestly. Which brings us back to Nishida, whose lifetime philosophical project sprang from and kept returning to one central concern: what it means to achieve an unmediated communion with reality or, as he put it, to 'know facts just as they are'.[333]

What it means, according to Nishida, is that we discover the cascade-like, self-enveloping, and self-reflecting nature of reality. True, in his enthusiasm to capture this nature of reality, Nishida presents us with a dauntingly complex metaphysical system: his 'logic of *basho*', comprising at least three major *bashos* and multiple (up to nine, depending how you count them)[334] gradations or degrees of reality within its concentric structure. In Nishida's defence, and at the risk of stating the obvious, I suggest that the complexity of his system reflects the complexity of what he is trying to do, namely to convey the unfamiliar nature of 'true' reality at the same time as making room for the more familiar (and, he would say, 'adulterated': my page 70) ways in which we experience the world. While the *basho* of beings (the least real *basho*) makes room for individual beings, everything that exists up to and including the universe, it is the *basho* of consciousness (a sort of transitional, semi-real *basho*) that makes room for our knowledge of and judgements about the individual beings. To know and judge the world we have to be conscious of it, and we are conscious of it (or more correctly, the world is something we are conscious of: page 77) only because the *basho* of beings in which it is enveloped is itself enveloped by the *basho* (or 'field') of consciousness. All very complicated, but maybe an unavoidable consequence of the 'logic' of Nishida's 'logic of *basho*'.

In turn, the *basho* of consciousness is enveloped by the *basho* of true nothingness, the most real *basho* and the most original and important aspect of Nishida's metaphysical system.[335] The *basho* of true nothing 'sees' itself, *within itself*, mirrored in 'everything that is and everything that is at work'.[336] But it does not cause, 'possess', or act upon the beings or anything else it envelops. It lets them be. Likewise, it encompasses all apparent opposites, contradictions and dualities, yet does so without resolving or reconciling them, for it 'transcends the opposition of being and nothing in every sense and enables them to be established within itself.'[337] Such, says Nishida, is the true nature of reality, which is as it is because the *basho* of true nothing is indeed a nothing, an absolute nothing. And that, we can't help noticing, recalls a conception of reality, encountered several times throughout this essay, which is both very old and relatively new. The old one is the great nothingness of Buddhism, *śūnyatā*, itself without time, space or differentiation of any kind and yet the origin of time, space, and every other differentiation in the world as we experience it. The newer conception is the Big Bang theory, currently our best scientific account of how the universe came into being 13.8 billion years ago: out of nothing, spontaneously, discontinuously, acausally. In Nishida, as in Buddhism, there is no Big Bang instant at which matter and energy begin. Yet in all three — Nishida, Buddhism and the Big Bang theory — there is the same unity (or shall we call it *interchangeability*?) of a profusely differentiated world and an absolutely undifferentiated nothingness.

To complete our definition of 'of the world', Nishida suggests, we must add that we are also of the nothing: not in Heidegger's anthropocentric sense that we are *nothing but* the possibilities and understanding determined for us by the 'they', rather in Nishida's metaphysical sense that we are one with the nothingness of a seeing that has no seer.

To know all this, according to Nishida, is to know that 'the self and the universe share the same foundation; or rather, they are the same thing.'[338] Moreover, it achieves the goal of 'true religion' (or religion as 'the consummation of philosophy'), which is to say, 'the transformation of the self and the reformation of life'.[339] Now we are bound to ask of Nishida's transformation the same question we asked earlier of Heidegger's *Augenblick* (the transformative moment in which Dasein is open to resoluteness)[340] — what practical difference, if any, could or should it make to the way we live our lives? With Nishida as with Heidegger, it does not necessarily follow that anything would be different in practical terms. Just as the day-to-day life of an authentic Dasein is barely distinguishable from that of an inauthentic or 'undifferentiated' Dasein, it is not at all obvious how the everyday practicalities of a transformed, 'true' self would change. On the contrary, by alluding to the Chan master Linji (Rinzai), who taught that 'You have only to be ordinary, ... defaecating, urinating, dressing, eating, and lying down to rest', Nishida makes a positive virtue of the ordinariness of daily life or *byōjōtei*.[341] Evidently it is not his intent, any more than it is Heidegger's, to encourage us to abandon our familiar patterns of life, still less to renounce the world altogether. It seems there is nothing much to choose between Heidegger's supposedly joyful liberation and Nishida's 'reformation of life'. Is it a matter of taste which we prefer?

The answer, I think, is that it only seems a matter of taste because we are asking the wrong question, or rather, the right question in the wrong way. 'Liberation' and the 'reformation of life' are not primarily or directly about the consequences of a transformation in our understanding of the world but about the character or quality of the transformation itself — the character or quality that Heidegger calls 'resoluteness' and Nishida calls 'the good'. As soon as the question is recast in these terms, there is every reason to choose Nishida over Heidegger. At best, Heidegger's resoluteness offers us the prospect of an ambiguous and fragile detachment from the they-world, yet no prospect of a meaningful relation with the world itself (in the sense of everything that exists or, to borrow Heidegger's own phrase one last time, *what is*). The character or quality of what Nishida offers is profoundly different. He defines our 'greatest good' as 'unit[ing] with the true reality of the self' and displaying our 'innate nature — just as a bamboo ... or a pine tree fully displays its nature'. [342] That is, our greatest good is to express our *suchness*, our condition of being just as we are and not otherwise, which, if only we knew it, is manifest in the ordinariness of our daily lives. The possibility of every conceivable thing or being in the world, every possibility that is, has been, and might ever be, comes out of the nothing. [343] Yet as conscious human beings we have a unique ability to acknowledge that, of all those possibilities, it is *ours* that embody *our* just-as-it-isness. And we have a responsibility to do so, precisely because we are uniquely capable of grasping the nature of reality and that is the basis on which 'the good' has to be understood. [344] I believe this notion of responsibility is implicit in Nishida's '[t]he laws of morality ... come to be included in the laws of reality'. [345] In other words, we have not just a responsibility but a moral responsibility to acknowledge our suchness.

The 'natural' knowledge I propose has, then, three main elements. First is the recognition that there is no reason, necessity or purpose for the world to be as it is. The hard evidence that demands the recognition is the century's worth of cosmology and physics that now robustly supports the Big Bang theory and makes it, in effect, the true creation story. Of the figurative creation stories, Lucretius's didactic poem is the most honest because it flatly denies any role for divine agency — no place for the God of Genesis or Plato's creator god — and insists upon the sheer randomness of the process by which the universe came into being.

The second element consists in accepting that the world has no obligation to be intelligible to us. In their different ways, and implicitly if not explicitly, Lucretius, Heidegger and Nishida all agree on this. However, I stand by my view that the most persuasive advocates of unintelligibility are the physicist-philosophers Bohr and d'Espagnat. Bohr calls for 'a radical revision of our attitude towards the problem of physical reality', while d'Espagnat reprimands his peers for teaching and perpetuating the 'falsehoods' of a 'crude model [of the world] tailored to specific purposes of ours'.[346] What they are both saying, d'Espagnat more stridently than Bohr, is that we must accept reality on its own terms — accept, and teach it, even if it seems incomplete, inconvenient, unreasonable or otherwise fails to 'make sense' in our terms.

The third element starts with the recognition that we are *of* as well as *in* the world. As far as it relates to the human world, it is comprehensively expressed by Heidegger's concept of 'Being-already-amidst-the-world'. But as it relates to the world of everything that is, the whole universe, it is best expressed by Nishida when he says to know that 'the self and the universe share the same foundation'[347] is to know that we are both of the world and of the nothingness that envelops the world.

As I said at the outset, I am not going to end with an attempt at prescribing what difference, practical or otherwise, 'natural' knowledge should make to the way we live our lives. Any difference it does make is surely a matter of individual conscience and choice, which, by definition, is not something that can be prescribed. What I have tried to do, hopefully with at least some success, is to identify the character or quality of the transformation in our understanding of the world that comes with 'natural' knowledge. The rest is up to us — or, if I may say so, up to you.

Of all the philosophers, physicists, and others I have chosen to bring together in this essay, Nishida gets us closest to *unseeing* the world as we think we know it, the better to see it anew. Although his earnest project to coin a language to convey the non-verbal insights of Chan/Zen Buddhism is at best a partial success, there are those brief moments when his project finds its own voice, simpler and quieter, often figurative, issuing more directly from — and speaking to — intuition, not intellect:

> We reach the quintessence of good conduct only when subject and object merge, self and things forget each other, and all that exists is the activity of the sole reality of the universe. At that point we can say that things move the self or that the self moves things, that Sesshū painted nature or that nature painted itself through Sesshū. There is no fundamental distinction between things and the self, for just as the objective world is a reflection of the self, so is the self a reflection of the objective world. The self does not exist apart from the world that it sees.[348]

As here with his reference to Sesshū, Nishida's imagery tends to come from ink painting and calligraphy. If I had to pick out just one example, it would be a fleeting image from his essay of 1926:

> Drawn in the space of true nothing, even a single point or a single brushstroke is a living reality.[349]

The part is in the whole and the whole is in the part. Like the point or brushstroke, we are one with nothing and yet, in our suchness, we are a unique part of everything. This, Nishida says, is the 'original state of independent, self-sufficient consciousness' in which we 'know facts just as they are'.[350] Or as I have suggested, it is the state of 'natural' knowledge in which we regain to know aright and can hope to reach a completely honest, secular, and long overdue accommodation with the world.

And if now is not the time to reach it, when will be?

# Sources

**Abe, Masao (1988)**. 'Nishida's Philosophy of "Place"' in *International Philosophical Quarterly* Vol. XXVIII, No. 4, Issue 112 (December 1988), pp. 355-71. Translation by Christopher Ives.

—— **(1997)**. 'Śūnyatā as Formless Form: Plato and Mahāyāna Buddhism' in Steven Heine, editor, *Zen and Comparative Studies* (London: MacMillan Press, 1997), pp. 139-48.

**Aspect, Alain et al. (1982)**. 'Experimental Test of Bell's Inequalities Using Time-Varying Analyzers' in *Physical Review Letters* Vol. 49, No. 25 (20 December 1982), pp. 1804-07.

**Awakawa, Yasuichi (1971)**. *Zen Painting* (Tokyo: Kodansha International, 1971). Translated by John Bester.

**Bacon, Francis (c. 1603)**. *Valerius Terminus of the Interpretation of Nature*. Text available online at Project Gutenberg (www.gutenberg.org/files/3290/3290-h/3290-h.htm).

**Bergquist, J.C. et al (1986)**. 'Observation of Quantum Jumps in a Single Atom' in *Physical Review Letters* Vol. 57, No. 14 (6 October 1986), pp. 1699-1702.

**Bohr, Niels (1928)**. 'The Quantum Postulate and the Recent Development of Atomic Theory' in *Nature* Vol. 121, No. 3050 (14 April 1928), pp. 580-90.

—— **(1935)**. [Reply to] 'Can Quantum-Mechanical Description of Physical Reality Be Considered Complete?' in Toulmin (1970), pp. 130-42.

—— **(1949)**. 'Discussion with Einstein on Epistemological Problems in Atomic Physics' in Schilpp (1970), pp. 201-41.

—— **(1962)**. Interview with Thomas Kuhn, Leon Rosenfeld, Aage Petersen and Erik Rudinger. The interview, conducted in Copenhagen on 17 November 1962 and now part of the Niels Bohr Library & Archives of the American Institute of Physics, is available online (aip.org/history-programs/niels-bohr-library/oral-histories/4517-5).

**Born, Max (2005)**. *The Born-Einstein Letters 1916-1955: Friendship, Politics and Physics in Uncertain Times* (New York: Macmillan, 2005). Translated by Irene Born.

**Campbell, Gordon, editor (1990)**. *John Milton: Complete English Poems, Of Education, Areopagitica* (London: Dent/Everyman, 1990).

**Carter, Robert (1997)**. *The Nothingness Beyond God: An Introduction to the Philosophy of Nishida Kitarō* (St Paul, MN: Paragon House Publishers, 1997).

**Cohen, S.M., Patricia Curd and C.D.C. Reeve, editors (2000)**. *Readings in Ancient Greek Philosophy* (Indianapolis: Hackett Publishing, 2000).

**Deprit, Andre (1984)**. 'Monsignor Georges Lemaître' in A. Berger, editor, *The Big Bang and Georges Lemaître* (Dordrecht: Springer-Verlag, 1984), pp. 363-92.

**d'Espagnat, Bernard (1972)**. 'Quantum Logic and Non-Separability' in J. Mehra, editor, *The Physicist's Conception of Nature* (Dordrecht: D. Reidel, 1973), pp. 714-35.

—— **(1979)**. 'The Quantum Theory and Reality' in *Scientific American* Vol. 241, No. 5 (November 1979), pp. 128-40.

**Dilworth, David (1987)**. 'Nishida's Critique of the Religious Consciousness' in *Last Writings: Nothingness and the Religious Worldview* (Honolulu: University of Hawai'i Press, 1987), pp. 1-45. Dilworth's introduction to his own translation of Nishida's '*Bashoteki ronri to shūkyōteki sekaikan*' [西田 (1945) below].

**Dreyfus, Hubert (1987)**. 'Husserl, Heidegger and Modern Existentialism' in Bryan Magee, editor, *The Great Philosophers: An Introduction to Western Philosophy* (Oxford: Oxford University Press, 1988), pp. 254-77.

—— **(1991)**. *Being-in-the-World: A Commentary on Heidegger's Being and Time, Division I* (Cambridge, MA: MIT Press, 1991).

**Einstein, Albert (1934)**. 'On Scientific Truth' in *Essays in Science*, translated by Alan Harris (New York: Wisdom Library, 1934), p. 11.

—— **(1949)**. 'Autobiographical Notes' in Schilpp (1970), pp. 1-94. Translated by Schilpp.

—— **(2015)**. *Letters to Solovine 1906-1955* (New York: Philosophical Library, 2015).

**Einstein, Albert, B. Podolsky and N. Rosen (1935)**. 'Can Quantum-Mechanical Description of Physical Reality Be Considered Complete?' in Toulmin (1970), pp. 123-30.

**Fine, Arthur (2007)**. 'Bohr's Response to EPR: Criticism and Defense' in *Iyyun • The Jerusalem Philosophical Quarterly* Vol. 56 (January 2007), pp. 31-56.

—— (2017). 'The Einstein-Podolsky-Rosen Argument in Quantum Theory' in *The Stanford Encyclopedia of Philosophy* (at https://plato.stanford.edu/archives/win2017/entries/qt-epr/). Winter 2017 version, edited by Edward Zalta.

**Fletcher, William, translator (1886)**. Lactantius, *Divine Institutes*. Available online (www.newadvent.org/fathers/07013.htm).

**Gamow, George (1970)**. *My World Line: An Informal Autobiography* (New York: Viking Press, 1970).

**Gleick, James (1986)**. 'Physicists Finally Get to See Quantum Jump With Own Eyes' in *The New York Times* of 21 October 1986; *Science Times Supplement*, pp. C1 and C4.

**Greenblatt, Stephen (2012)**. *The Swerve: How the Renaissance Began* (London: Vintage Books, 2012).

**Gribbin, John (1992)**. *In Search of Schrödinger's Cat* (London: Black Swan, 1992).

**Halvorson, Hans and R. Clifton (2001)**. 'Reconsidering Bohr's Reply to EPR' in the open-access physics archive *arXiv* (arXiv:quant-ph/0110107v1 17 Oct 2001).

**Hawking, Stephen (1989)**. *A Brief History of Time: From the Big Bang to Black Holes* (New York: Bantam Books, 1989).

**Heidegger, Martin (1927)**. *Being and Time*. Translation of *Sein und Zeit* by John Macquarrie and Edward Robinson (Oxford: Blackwell, 2003).

—— (1929). *What is Metaphysics?* David Krell's translation of *Was ist Metaphysik?* in D.F. Krell, editor, *Martin Heidegger: Basic Writings* (Abingdon, Oxon: Routledge, 2011), pp. 45-57.

—— (1943). The postscript Heidegger appended to *Was ist Metaphysik?* in 1943, translated by R.F.C. Hull and Alan Crick. Reprinted in Kaufmann (1975), pp. 257-64.

—— (1949). 'The Way Back into the Ground of Metaphysics'. Translation by Walter Kaufmann of '*Der Rückgang in den Grund der Metaphysik*', the long introduction Heidegger added to *Was ist Metaphysik?* in 1949. In Kaufmann (1975), pp. 265-79.

**Heisenberg, Werner (1927)**. 'The Physical Content of Quantum Kinematics and Mechanics' in John Wheeler and Hubert Zurek, editors, *Quantum Theory and Measurement* (Princeton: Princeton University Press, 1983), pp. 62-84. Translation by Wheeler and Zurek.

**Heisig, James (2001)**. *Philosophers of Nothingness: An Essay on the Kyoto School* (Honolulu: University of Hawai'i Press, 2001).

**Heisig, James and John Maraldo, editors (1995)**. *Rude Awakenings: Zen, the Kyoto School, & the Question of Nationalism* (Honolulu: University of Hawai'i Press, 1995).

**Inwood, Brad and Lloyd Gerson, translators/editors (1994)**. *The Epicurus Reader: Selected Writings and Testimonia* (Indianapolis: Hackett Publishing, 1994).

**Jacques, Vincent et al. (2007)**. 'Experimental Realization of Wheeler's Delayed-Choice Gedanken Experiment' in *Science* Vol. 315, No. 5814 (16 February 2007), pp. 966-68.

**Kaufmann, Walter, editor (1975)**. *Existentialism from Dostoevsky to Sartre* (New York: Plume Books, 1975).

**Krauss, Lawrence (2012)**. *A Universe from Nothing: Why There Is Something Rather than Nothing* (London: Simon & Schuster, 2012).

**Latham, R.E., translator (2005)**. Lucretius, *On the Nature of the Universe* (London: Penguin Books, 2005). Latham's prose translation of 1951, revised by John Godwin.

**Lee, H.D.P., translator (1977)**. Translation of Plato's *Timaeus* in *Timaeus and Critias* (Harmonsdworth: Penguin Classics, 1977).

**Lemaître, Georges (1927)**. 'A Homogeneous Universe of Constant Mass and Increasing Radius accounting for the Radial Velocity of Extra-galactic Nebulæ' in *Monthly Notices of the Royal Astronomical Society* Vol. 91, Issue 5 (March 1931), pp. 483-90.

—— **(1931)**. 'The Beginning of the World from the Point of View of Quantum Theory' in *Nature* Vol. 127, No. 3210 (9 May 1931), p. 706.

**Lovejoy, Arthur (1964)**. *The Great Chain of Being: A Study of the History of an Idea* (Cambridge, MA: Harvard University Press, 1964).

**MacKenna, Stephen and B.S. Page, translators (1988)**. *Plotinus: The Six Enneads* (Chicago: University of Chicago/Encyclopædia Britannica Great Books of the Western World, 1988).

**Marletto, Chiara et al. (2017)**. 'Entanglement between living bacteria and quantized light witnessed by Rabi splitting' in *arXiv* (arXiv:1702.08075 [quant-ph]).

**McRae, John, translator (2000)**. *The Platform Sutra of the Sixth Patriarch* (Berkeley, CA: Numata Center for Buddhist Translation and Research, 2000). From the Chinese, available online (at http://www.thezensite.com./ZenTeachings/Translations/PlatformSutra_McRaeTranslation.pdf).

**Morris, Mary and G.H.R. Parkinson, translators (1984)**. *Leibniz: Philosophical Writings* (London: Dent/Everyman, 1984).

**Nagel, Thomas (2012)**. *Mind and Cosmos: Why the Materialist Neo-Darwinian Conception of Nature Is Almost Certainly False* (Oxford: Oxford University Press, 2012).

**Nagourney, Warren et al. (1986)**. 'Shelved Optical Electron Amplifier: Observation of Quantum Jumps' in *Physical Review Letters* Vol. 56, No. 26 (30 June 1986), pp. 2797-99.

**Nishida, Kitarō (1911)**. *Zen no kenkyū* [西田 (1911) below], translated as *An Inquiry into the Good* by M. Abe and C. Ives (New Haven: Yale University Press, 1990).

—— **(1926)**. '*Basho*' [西田 (1926)], translated as 'Basho' by John Krummel and S. Nagatomo in *Place and Dialectic: Two Essays by Nishida Kitarō* (New York: Oxford University Press, 2012), pp. 49-102.

—— **(1928)**. '*Eichiteki sekai*' [西田 (1928)], translated as 'The Intelligible World' by Robert Schinzinger in *Intelligibility and the Philosophy of Nothingness: Three Philosophical Essays* (Tokyo: Maruzen, 1966), pp. 69-141.

—— **(1945)**. '*Bashoteki ronri to shūkyōteki sekaikan*' [西田 (1945)], translated as 'The Logic of *Topos* and the Religious Worldview' by M. Yusa. In two parts in *The Eastern Buddhist* New Series Vol. 19, No. 2 (Autumn 1986), pp. 1-29; and Vol. 20, No. 1 (Spring 1987), pp. 81-119.

**西田幾多郎著 (1911)**。『善の研究』（岩波書店、2019）。

—— **(1926)**。「場所」、上田閑照編『西田幾多郎哲学論集Ⅰ』（岩波書店、2019）、67-151頁。

—— **(1927)**。「働くものから見るものへ」、序、上田閑照編『西田幾多郎哲学論集Ⅰ』（岩波書店、2019）、33-36頁。

—— **(1928)**。「叡知的世界」、上田閑照編『西田幾多郎哲学論集Ⅰ』（岩波書店、2019）、189-253頁。

—— **(1945)**。「場所的論理と宗教的世界観」、上田閑照編『西田幾多郎哲学論集Ⅲ』（岩波書店、2019）、299-397頁。

**O'Raifeartaigh, Cormach and Simon Mitton (2018)**. 'Interrogating the Legend of Einstein's "Biggest Blunder"' in *Physics in Perspective* Vol. 20, Issue 4 (December 2018), pp. 318-41.

**Pelikan, Jaroslav, editor (1958)**. *Luther's Works, Volume 1: Lectures on Genesis, Chapters 1-5* (St Louis, MO: Concordia Publishing, 1958).

**Petersen, Aage (1963)**. 'The Philosophy of Niels Bohr' in A.P. French and P.J. Kennedy, editors, *Niels Bohr: A Centenary Volume* (Cambridge, MA: Harvard University Press, 1985), pp. 299-310.

**Popper, Karl (1982)**. 'A Metaphysical Epilogue' in W.W Bartley, editor, *Quantum Theory and the Schism in Physics* (London: Hutchinson, 1982), pp. 159-211.

**Rosenfeld, Leon (1963)**. 'Niels Bohr's contribution to epistemology' in *Physics Today* Vol. 16, No. 10 (October 1963), pp. 47-54.

**Rovelli, Carlo (2017)**. *Reality Is Not What It Seems*. Translation of *La realtà non è come ci appare* by Simon Carnell and Erica Segre (London: Penguin Books, 2017).

**Sartre, Jean-Paul (1938)**. *La nausée*, translated as *Nausea* by Robert Baldick (Harmondsworth: Penguin Books, 1971).

────── **(1943)**. *L'être et le néant: Essai d'ontologie phénoménologique*, translated as *Being and Nothingness: An Essay on Phenomenological Ontology* by Hazel Barnes (London: Routledge, 1993).

────── **(1946)**. *Réflexions sur la question juive*, translated as *Anti-Semite and Jew* by George Becker (New York: Schocken Books, 1995). An expanded version of Sartre's *'Portrait de l'antisémite'*, originally published in the journal *Les Temps modernes* in December 1945.

**Sasaki, Ruth, translator (2009)**. *The Record of Linji* (Honolulu: University of Hawai'i Press, 2009). Translation of the *Linji lu*, sayings of the Chan priest Linji, also known as Rinzai in the Japanese pronunciation of his name. Edited by Thomas Kirchner.

**Sauter, T. et al. (1986)**. 'Observation of Quantum Jumps' in *Physical Review Letters* Vol. 57, No. 14 (6 October 1986), pp. 1696-98.

**Schilpp, P.A., editor (1970)**. *Albert Einstein: Philosopher-Scientist* (La Salle, Illinois: Open Court/Library of Living Philosophers, 1970). First published in 1949 to celebrate Einstein's seventieth birthday.

**Stadermann, H.K.E. et al. (2019)**. 'Analysis of secondary school quantum physics curricula of 15 different countries: Different perspectives on a challenging topic' in *Physical Review Physics Education Research* Vol. 15, 010130 (22 May 2019: 10.1103/PhysRevPhysEducRes.15.010130).

**Toulmin, Stephen, editor (1970)**. *Physical Reality: Philosophical Essays on Twentieth-Century Physics* (New York: Harper Torchbooks, 1970).

**Ueda, Shizuteru (1982)**. 'Emptiness and Fullness: Śūnyatā in Mahāyāna Buddhism' in *The Eastern Buddhist* Vol. 15, No. 1 (Spring 1982), pp. 9-37. Translated by James Heisig and Frederick Greiner.

────── **(1995)**. 'The Difficulty of Understanding Nishida's Philosophy' in *The Eastern Buddhist* Vol. 28, No. 2 (Autumn 1995), pp. 175-82. Translated by Thomas Kirchner.

**Waldenfels, Hans (1980)**. *Absolute Nothingness: Foundations for a Buddhist-Christian Dialogue* (Nagoya: Chisokudō Publications, 2020). Translated from German by James Heisig.

**Warnock, Mary (1992)**. *Existentialism* (Oxford: Oxford University Press, 1992).

**Weiss, Achim (2006)**. 'Elements of the past: Big Bang Nucleosynthesis and observation' in *Einstein Online* Band 02 (2006), 02-1019 (einstein-online.info/en/spotlight/bbn_obs/).

**Wheeler, John (1979)**. 'Beyond the Black Hole' in Harry Woolf, editor, *Some Strangeness in the Proportion: A Centennial Symposium to Celebrate the Achievements of Albert Einstein* (Reading, MA: Addison-Wesley Publishing, 1980), pp. 341-75.

**Whitaker, M.A.B. (2004)**. 'The EPR Paper and Bohr's Response: A Re-Assessment' in *Foundations of Physics* Vol. 34, No. 9 (September 2004), pp. 1305-40.

**Wilkinson, Geoffrey (2012)**. *Certainty, that thing of indefinite approximation: a quest through lives and literatures* (Franksbridge, Powys: Geoffrey M. Wilkinson, 2024 revised reprint).

────── **(2015)**. 'The Frog and the Basilisk' in *Comparative & Continental Philosophy* Vol. 7, No. 1 (May 2015), pp. 44-51. Revised and reprinted in Wilkinson (2019), pp. 41-52.

────── **(2019)**. *Going to the Pine: Four Essays on Bashō* (Franksbridge, Powys: Geoffrey M. Wilkinson, 2019).

**Yin, Juan et al. (2017)**. 'Satellite-based entanglement distribution over 1200 kilometers' in *Science* Vol. 356, Issue 6343 (16 June 2017), pp. 1140-44.

# Notes

p./pp. = source or my main text page reference/s
n./nn. = source or my own endnote reference/s

## 2. What we have lost

[1] Nagel (2012), p. 29.

[2] Ibid., p. 128.

[3] Ibid., p. 6.

[4] Ibid., p. 20.

[5] Ibid., p. 33. Elsewhere (pp. 92, 95 and 123) Nagel speaks variously of 'natural teleology', 'teleological naturalism', and the 'teleological hypothesis ... that [the existence of the genetic material natural selection works on] may be determined not merely by value-free chemistry and physics but also by ... a cosmic predisposition to the formation of life, consciousness, and the value that is inseparable from them.'

[6] Ibid., p. 44.

[7] *Timaeus* 28c-29b, 30c-31b and 39e; pp. 41, 43 and 54 in the translation by H.D.P. Lee in Lee (1977).

[8] *Timaeus* 29e; p. 42 in Lee (1977).

[9] *Fifth Ennead*, Tractate II.1; as translated by Arthur Lovejoy in Lovejoy (1964), p. 62. For an interpretation of Plotinus's One from a non-Western philosophical perspective, see my n. 233 below.

[10] *Fifth Ennead*, Tractate II.2; Lovejoy translation in Lovejoy (1964), p. 63. The world has not been created directly by the One, which is at the top of Plotinus's metaphysical hierarchy: the One has generated *Nous* (commonly translated as 'Intellect' or 'Intellectual-Principle'), and *Nous* in turn has generated Soul. It is Soul, driven by a procreative 'appetite' of its own, that has made the stars, the planets and everything else in the visible universe. In a passage that might be mistaken for Genesis but that Plotinus is speaking of Soul, not God, we are told that Soul 'is the author of all living things, ... it has breathed the life into them all, whatever is nourished by earth and sea, all the creatures of the air, the divine stars in the sky; it is the maker of the sun; itself formed and ordered this vast heaven and conducts all that rhythmic motion.' From the *Fifth Ennead*, Tractate I.2; p. 208 in the translation by Stephen MacKenna and B.S. Page in MacKenna and Page (1988).

[11] The subject of Lovejoy's wonderful book, which is titled *The Great Chain of Being*.

[12] From *On the Ultimate Origination of Things* (*De rerum originatione radicali*), translated by Mary Morris and G.H.R. Parkinson in Morris and Parkinson (1984), p. 136.

# Notes

[13] Ibid., p. 139.

[14] Ibid., p. 137.

[15] Ibid., p. 138.

[16] From Leibniz's *Théodicée*, published in 1710; Lovejoy translation in Lovejoy (1964), pp. 166-67.

[17] It is very striking that Leibniz's insistence that God does not play games of chance prefigures so exactly Albert Einstein's protestation that he could not believe God played dice with nature (albeit Einstein's protestation may not have had the significance it is generally accorded: see my n. 90 below).

[18] Even to entertain the notion that we do inhabit a world of blind chance would mean, in Lovejoy's memorable phrase, 'placing Caprice on the throne of the universe.... It [would imply] that Nature, having no determining reason in it, flouts and baffles the reason that is in man. A world where chance-happening had so much as a foothold would have no stability or trustworthiness; uncertainty would infect the whole.' Lovejoy (1964), p. 168.

[19] From *New Essays on the Human Understanding* (*Nouveaux essais sur l'entendement humain*); p. 158 in Morris and Parkinson (1984). Italics reproduced as in their translation. The *New Essays*, a counterblast to John Locke's *Essay on the Human Understanding*, were finished in 1704 but not published until 1765.

[20] As I understand it, the three teams — at the University of Washington, Seattle, the University of Hamburg, and the National Bureau of Standards in Boulder, Colorado — were the first ever to observe and demonstrate the existence of quantum jumps. The microscope was improvised by the University of Washington team. My original source was Gleick (1986), which I came across, fittingly enough, by chance in a newspaper I happened to buy that day. For those with the technical knowledge to read them, the three peer-reviewed journal reports can be found in Nagourney et al. (1986), Sauter et al. (1986) and Bergquist et al. (1986) respectively.

[21] In 1900 Max Planck had suggested quantization as a means of explaining the odd behaviour of the radiation emitted by hot objects (the 'blackbody radiation problem') but he did so in a rather makeshift way and its importance was not immediately appreciated. Nonetheless, Planck was belatedly awarded the Nobel Prize in Physics in 1918 for this work. Einstein was one of the first, if not the first, to see quantization as more than a mathematical sleight of hand and explore its true significance — hence his crucial role in the early development of quantum theory.

[22] The physicist John Wheeler believed that Einstein's assumption of a static universe was heavily influenced by his reading of Spinoza. In turn, Wheeler credited the theologian Hans Küng for this insight: Wheeler (1979), pp. 352-54 and his nn. 83, 84 on p. 374.

# Notes

[23] The main source for Einstein's 'biggest blunder' comment is a single paragraph in the autobiography of the physicist George Gamow: Gamow (1970), p. 44. As there is nothing in Einstein's own papers or correspondence to corroborate Gamow's account, it is quite widely regarded as apocryphal. But research by Cormac O'Raifeartaigh and Simon Mitton suggests it is very plausible that Einstein would have made such a comment, which is consistent with his known sentiments at the time: O'Raifeartaigh and Mitton (2018).

[24] After serving in a Belgian artillery regiment during the First World War, Lemaître read physics and mathematics while training for the priesthood, and then moved into cosmology. From 1923 he studied in Cambridge with Sir Arthur Eddington, and latterly at the Massachusetts Institute of Technology. His 1927 paper was published in the journal *Annales de la Société scientifique de Bruxelles*: my citation, Lemaître (1927), is the abridged English translation with additional commentary by Lemaître that appeared in 1931. Einstein himself is said to have responded to the paper, which Lemaître had presented to him while attending the Solvay Conference on Physics in Brussels in October 1927, with the barbed compliment 'Your calculations are correct, but your physics is abominable [*Vos calculs sont corrects, mais votre physique est abominable*]': Deprit (1984), p. 370. Unbeknown to Lemaître, in 1922 the Russian physicist and mathematician Alexander Friedmann had independently derived an expanding-universe model from general relativity, also retaining Einstein's cosmological constant. In the sense that both Friedmann and Lemaître recognized that the universe is expanding, and therefore must have had a beginning in the past, I suppose that both should be credited with introducing the notion of a Big Bang into cosmology. It is striking, though, that Lemaître came especially close to anticipating our current understanding of the Big Bang, even to the point of invoking quantum theory. In 1931, for instance, he suggested that 'If we go back in the course of time we must find fewer and fewer quanta, until we find all the energy of the universe packed in a few or even in a unique quantum…. [I]t may be that an atomic nucleus must be counted as a unique quantum…. [If so,] we could conceive the beginning of the universe in the form of a unique atom, the atomic weight of which is the total mass of the universe': Lemaître (1931).

[25] It could be argued that Einstein's mistake was not to have added the cosmological constant in 1917 but to have removed it in 1931, for current thinking today is that the accelerating expansion of the universe may well be driven by the dark energy, which — another irony — can be described using general relativity with a cosmological constant. In 1998 the value of the cosmological constant was measured by the astronomers Saul Perlmutter, Brian Schmidt and Adam Riess, work for which they were awarded the Nobel Prize for Physics in 2011.

[26] Leavitt was a human 'computer' at the Harvard College Observatory, responsible for cataloguing the brightness of stars detected on photographic plates. The brightness of Cepheid variable stars varies over some regular period. Leavitt's discovery in 1912 of the relation between the brightness and period of variation enabled her to use the brightness as a benchmark for

# Notes

measuring the distance to Cepheid variables and, by extension, to their host galaxies.

[27] In observations first begun in 1912, Slipher had detected that the light from most of the galaxies he studied was red-shifted. Humason was a key member of the technical staff at the Mt Wilson Observatory near Los Angeles where Hubble made his own observations.

[28] At the time of writing, there are unresolved discrepancies in the value of the Hubble constant: some observations of the movements of galaxies suggest a velocity above 70 kilometres per second, while other types of measurement (see my n. 34) come in below 70 kilometres per second. Hubble was way out in his own calculation of the recession velocity, estimating it at 500 kilometres per second (which was awkward because it implied an age for the universe of about 1.5 billion years, whereas radiometric measurements of geological samples were already indicating that the *earth* was about 3 billion years old!).

[29] Krauss (2012), p. 18 (relates to the figure on his p. 111). His italics. Observations to test this and other predictions involve analysis of the emission lines (telltale lines in a spectrum of electromagnetic radiation corresponding to the discrete wavelengths of known elements) detected in galaxies relatively close to our own. The purpose of the analysis is, in effect, to factor out the proportion of elements (especially oxygen and nitrogen) that would have formed through stellar nuclear processes, such that those elements which are left could only have formed in the Big Bang nucleosynthesis. Achim Weiss of the Max Planck Institute for Astrophysics in Germany has a detailed but accessible explanation of this analysis: Weiss (2006).

[30] The reason the picture stops 380,000 years short of the Big Bang itself is that before then the universe was mostly filled with a dense plasma of charged particles, opaque to photons and therefore invisible to photon-dependent observation. At 380,000 years, the temperature of the universe would have been about 3,000 kelvins. To the best of my understanding, the anisotropies indicate the structures in the background radiation (which in turn may reflect quantum fluctuations or 'ripples' at some instant of the Big Bang) that gave rise to the stars and galaxies as the universe expanded.

[31] Hawking (1989), pp. 9, 50 and 114.

[32] Revised estimates of the age of the universe published by the Planck Collaboration in 2018 vary somewhat depending on different combinations of parameters: I understand that $13.797 \pm 0.023$ billion years is a baseline result. Full technical data can be found under 'Planck 2018 Results' at Planck Publications (www.cosmos.esa.int/web/planck/publications).

[33] The WMAP (Wilkinson Microwave Anisotropy Probe), launched in 2001, was developed jointly by the NASA Goddard Space Flight Center and Princeton University: follow the links on the WMAP Project website (http://wmap.gsfc.nasa.gov) for technical data.

# Notes

[34] There is especially good agreement between values for the Hubble constant: the Altacama measurements suggest a Hubble constant of 67.6 kilometres per second per megaparsec (67.6 ± 1.1 km/s/Mpc), a close match for the 67.3 kilometres per second (67.3 ± 0.6 km/s/Mpc) reported by the Planck Collaboration. The Atacama Cosmology Telescope (ACT) is an international collaboration supported by the US National Science Foundation: preprints of the 2020 research reports are available on the ACT website (act.princeton.edu/publications). See also n. 28 above.

[35] In Campbell (1990), p. 557.

[36] Ibid., p. 564.

[37] Genesis 2:19.

[38] From Luther's lecture on Genesis 2:19; translated by George Schick in Pelikan (1958), p. 121. Luther argued that Adam's naming of the beasts also accounted for his command over them, such that 'by one single word he was able to compel lions, bears, boars, tigers ... to carry out whatever suited their nature.' The text of the lecture, which Luther probably gave in Wittenberg in June 1535, survives only as transcribed and reworked by later editors, and hence there are authenticity issues with it.

[39] This wording comes from Bacon's *Valerius Terminus of the Interpretation of Nature*, an unfinished work thought to date to about 1603: Bacon (c. 1603), Chap. 1, para. 7. Although very close to the wording in Book One, I(3) of Bacon's *Of the Proficience and Advancement of Learning, Divine and Human*, published in 1605, personally I like the freshness of the earlier manuscript.

## 3. The fortuitous concourse of all things

[40] *On the Nature of the Universe*, Book 5.417-25; p 139 in Latham (2005).

[41] Ibid,.Book 5.435-36; p. 139 in Latham.

[42] Here I have conflated passages from Books 5.156-66 and 2.167-71; pp. 133 and 42 in Latham.

[43] Although I focus on Lucretius and Epicurus, this section is largely inspired by Carlo Rovelli's speculation that 'Perhaps, if all of the works of Democritus had survived, and nothing of Aristotle's, the intellectual history of our civilization would have been better....' Rovelli (2017), p. 20.

[44] Epicurus holds that the gods are immortal because they somehow regenerate themselves instead of dissolving like all other compound matter. *Letter to Menoeceus* 123; p. 28 in Inwood and Gerson (1994).

[45] *On the Nature of the Universe*, Book 2.1058-61; p. 64 in Latham.

[46] *Letter to Menoeceus* 134; p. 31 in Inwood and Gerson.

[47] *On the Nature of the Universe*, Book 2.251-60; p. 44 in Latham.

# Notes

[48] A saying attributed to Epicurus by the Greek Neoplatonist philosopher Porphyry, who lived in the third century AD. Text 124, p. 99, in Inwood and Gerson.

[49] *On the Nature of the Universe*, Book 4.10-24; p. 95 in Latham.

[50] Ibid., Book 2.646-48; p. 53 in Latham.

[51] Ibid., Book 5.1203; p. 159 in Latham. Lucretius is contrasting 'true' piety with the false piety, motivated by fear of the gods, of abject self-abasement in prayer, animal sacrifice, and other rituals.

[52] Ibid., Book 3.322; p. 75 in Latham.

[53] *Letter to Menoeceus* 135; p. 31 in Inwood and Gerson.

[54] *On the Nature of the Universe*, Book 5.8; p. 129 in Latham.

[55] *Letter to Menoeceus* 122; p. 28 in Inwood and Gerson.

[56] Ibid. 131-32; pp. 30-31 in Inwood and Gerson.

[57] Appetites exemplified elsewhere in literature by the grotesque figure of Sir Epicure Mammon in Ben Jonson's play *The Alchemist*, first performed in 1610. Is Sir Epicure, whose proposed menu includes 'the swelling unctuous paps / Of a fat pregnant sow, newly cut off' (II.ii.83-84), intended as a caricature of Epicurus? If so, Jonson appears to accuse Epicurus, perversely, of the boundless gluttony that in fact he denounced.

[58] *Letter to Menoeceus* 132; p. 31 in Inwood and Gerson.

[59] *On the Nature of the Universe*, Book 5.1117-19; p. 157 in Latham.

[60] A copy of the text was found at a monastery (possibly Fulda in what is now the modern state of Hesse) by Poggio Bracciolini, a papal secretary and avid hunter of ancient books. The fascinating story is told in Greenblatt (2012), to which I am heavily indebted.

[61] *On the Nature of the Universe,* Book 5.1129-30; p. 157 in Latham.

[62] This brief biography of Lucretius is included in Jerome's *Chronicle*, his Latin translation of the chronological tables compiled earlier in the fourth century AD by Eusebius, a Christian historian and polemicist. Eusebius's original Greek text is lost.

[63] From Lactantius's *Divine Institutes* Book 3, Chap. 17, translated by William Fletcher; quoted in Greenblatt, p. 102. The full text is available online at Fletcher (1886).

[64] *On the Nature of the Universe*, Book 2.1-12; p. 38 in Latham.

[65] Ibid., Book 5. 932 and 5.958-61; pp. 152-53 in Latham.

[66] Ibid., Book 5.1018-27; pp. 154-55 in Latham.

# Notes

[67] From the text now known as *On Choices and Avoidances*, generally attributed to Philodemus. Translation by Giovanni Indelli and Voula Tsouna-McKirahan; quoted in Greenblatt (2012), p. 77. Here virtually identical in wording to a passage in Epicurus's *Letter to Menoeceus* 132, it is one of the texts found in the so-called Villa of the Papyri in the ruins of Herculaneum, the coastal village destroyed together with Pompeii by the eruption of Mt Vesuvius in 79 AD. A striking metaphor: Philodemus as a voice of quiet reasonableness buried under dogma and disinformation for two thousand years, waiting to be rediscovered and heard anew.

## 4. No reasonable definition of reality

[68] Einstein (1949), p. 81.

[69] Bohr (1928), p. 580.

[70] Bohr (1949), pp. 201-02.

[71] Bohr's concern to avoid ambiguity is consistent with his belief that we are 'suspended in language' but that, with care, it is possible to 'extend' the conceptual frameworks on which we depend for 'unambiguous communication'. My source here is Aage Petersen, Bohr's assistant for 10 years from 1952: Petersen (1963), pp. 301-02. Frustratingly, it is not always clear whether Petersen is quoting Bohr verbatim, paraphrasing him, or interpreting him in his own Bohr-like terms. This is a serious flaw, I think, with regard to one passage in particular, "'There is no quantum world. There is only an abstract quantum physical description....'" (p. 305), which has been reprinted elsewhere and attributed without qualification to Bohr himself. Petersen's essay was first published in 1963, the year after Bohr's death.

[72] Heisenberg (1927), p. 83. I make no pretence of understanding the mathematical content of Heisenberg's paper, originally published in the journal *Zeitschrift für Physik* (Vol. 43, 1927).

[73] Just how far indeterminacy and other quantum effects extend into the macroscopic world is the subject of ongoing debate. Astonishingly, an experiment originally conducted in 2016 may have succeeded in entangling photons with photosynthetic molecules within live green-sulphur bacteria. While that evidence is inconclusive, a number of research groups propose to scale up further by working with tardigrades, aquatic creatures typically only about 0.5mm in size but orders of magnitude larger than bacteria. Analysis of the 2016 experiment, densely mathematical in places, can be found in Marletto et al. (2017).

[74] The EPR paper, Einstein, Podolsky and Rosen (1935), together with Bohr's reply, Bohr (1935), is reprinted in Toulmin (1970). The two papers first appeared in the journal *Physical Review* (Vols. 47 and 48, 1935).

[75] Page 124 in Toulmin. Italics as in the original text.

[76] Page 130 in Toulmin. Italics as in the original.

# Notes

[77] This is my understanding of Bohr (1935), especially the key passage dealing with what Bohr calls 'an ambiguity' in the EPR paper, reproduced on pp. 138-39 of Toulmin. My own wording and italics.

[78] Bohr (1949), p. 210. An expanded text of Bohr's 1927 lecture, which he gave at the International Congress of Physics in Como, Italy, appears in Bohr (1928).

[79] Strictly speaking, the experiments have tested Bell's inequality, a mathematical theorem which enables experimenters to interpret observed correlations (of spin rather than position/momentum) between entangled, spatially separated quantum systems. Formulated by the British physicist John Bell in 1964, the theorem makes it possible to evaluate the EPR thought-experiment against the Copenhagen interpretation because it shows that they predict different, and now testable, experimental outcomes. A result known as a violation of Bell's inequality is taken to endorse the predictions of quantum mechanics. There is a detailed, largely non-mathematical discussion of Bell's theorem and its implications in d'Espagnat (1979).

[80] The report on the third Paris experiment, which assumes technical and mathematical knowledge, is in Aspect et al. (1982).

[81] The 2017 experiment is both an impressive (and literal) demonstration of 'action at a distance' and suggests, as the Chinese researchers put it, 'the possibility of a future global quantum communication network'. Their findings, technical but mostly non-mathematical, are published in Yin et al. (2017).

[82] A similar criticism can be made against the EPR paper itself, which was written up by Podolsky. Soon after publication, Einstein, who apparently had not read the manuscript before its submission to *Physical Review*, complained that the argument was too abstract: as he put it in a letter to Erwin Schrödinger dated 19 June 1935, 'the essential thing was, so to speak, smothered by formalism.' My source is Fine (2017), §1.1; in §1.3 Fine describes Einstein's multiple attempts to clarify the EPR argument in later publications, none of which seem to have made any impression on Bohr.

[83] See Fine (2007), including pp. 4 ('recognizably positivist ... semantic doctrines'), 8 ('strong verificationism') and 24 ('Bohr begins with semantics, not physics').

[84] Halvorson and Clifton (2001), §1. The formalism of this paper is inaccessible to me, but I take the *classical* side of the 'dual requirements' to follow from Bohr's demand that measurements in a quantum-mechanical experiment be expressed in terms of classical, not quantum, physics. The authors conclude, §6, with a caveat: 'we wish to emphasize that Bohr is not so much concerned with what is *truly* real for the distant system [i.e., the second particle in the EPR thought-experiment] as he is with the question of what we would be *warranted in asserting* about the distant system from the standpoint of classical description.' All italics as in the original.

[85] Whitaker (2004), p. 1336. Whitaker asks whether two measurements (position *and* momentum) or just one (position *or* momentum) would be

# Notes

needed to confirm, or refute, the EPR argument that quantum mechanics is incomplete. Whitaker's own position is that 'there is no requirement even to discuss measurement of more than one quantity.' It is not clear, he says later, that Bohr's reply 'deals with the subtleties of the EPR argument. The possibility of making a hypothetical measurement of the position of one particle, according to EPR, establishes the fact that the other particle has a position after the measurement, therefore had it before the measurement, therefore has it even if we have no intention of making the measurement in the first place!' These quotes are from pp. 1320 and 1328.

[86] I leave it to the reader to decide whether my interpretation is supported or undermined by Bohr's own protestation — literally days before the EPR paper was sent for publication, and in language remarkably similar to that used in the EPR paper — that the notion of non-locality was 'completely incomprehensible' and that even to consider it would be to enter 'irrational territory'. He was discussing a different experimental setup, not the one envisaged in the EPR thought-experiment, but the issue was the same in both cases: 'action at a distance' in violation of the locality assumption. Bohr manuscript dated 21 March 1935, quoted in Fine (2007), p. 17, and Fine (2017), §2.

[87] Bohr (1935); p. 132 in Toulmin. Italics within the quoted passage are Bohr's.

[88] Another way of saying that Einstein was a scientific realist, i.e., one who holds that an objectively 'real' world exists independently of our knowledge of it. At the risk of another crude oversimplification, the realist position now tends to be associated with philosophers of science. Quantum physicists, on the other hand, tend towards a form of instrumentalism, the position (inelegantly paraphrased as 'Shut up and calculate') that quantum mechanics works well as a practical device for studying phenomena at the scale of Planck's constant, but tells us nothing about the 'reality' of the world at any scale. These positions represent two extremes in what Karl Popper called the 'schism' in physics, which he placed in the context of 'metaphysical research programmes' dating back to Pythagoras, Heraclitus and Parmenides: Popper (1982), pp. 159-77. See also nn. 89 and 90 following.

[89] Einstein (1934). Compare the wording of his letter to Maurice Solovine of 1 January 1951 in Einstein (2015), p. 105: 'I have no better expression than "religious" for confidence in the rational nature of reality insofar as it is accessible to human reason. Wherever this feeling is absent, science degenerates into uninspired empiricism.'

[90] I have chosen to focus on his complaint of incompleteness, but the usual perception is that Einstein was most offended by the implication that 'God plays dice', i.e., by the role of chance and indeterminism in quantum mechanics. It appears that even Max Born, who believed he had a good understanding of Einstein's thinking, misjudged Einstein's true concern and it was only in 1954, the year before Einstein died, that Born learnt from Wolfgang Pauli that 'Einstein's point of departure is "realistic" rather than

# Notes

"deterministic", which means that his philosophical prejudice is a different one.' Pauli letter to Born dated 31 March 1954 in Born (2005), p. 216.

## 5. Thrown into the world

[91] Translation by Walter Kaufmann in Heidegger (1943), p. 261. Mary Warnock and others have observed that in sentiment Heidegger here recalls a passage in one of Coleridge's essays: 'Hast thou ever raised thy mind to the consideration of existence, in and by itself, as the mere act of existing? Hast thou ever said to thyself thoughtfully *It is!* heedless in that moment whether it were a man before thee or a flower or a grain of sand ... without reference in short to this or that mode or form of existence? If thou hast attained to this thou wilt have felt the presence of a mystery which must have fixed thy spirit in awe and wonder.' Essay XI in *The Friend*, as quoted in Warnock (1992), p. 52.

[92] Translation by David Krell in Heidegger (1929), p. 49.

[93] Like many people, I imagine, I first got interested in Heidegger through 'Husserl, Heidegger and Modern Existentialism', a dialogue between the philosophers Hubert Dreyfus and Bryan Magee originally recorded by the BBC in 1987: Dreyfus (1987). I also rely heavily on the detailed commentary, which began life as Dreyfus's lecture notes on Division I of *Being and Time*, in Dreyfus (1991); heavily but not uncritically, I hope, and I am aware that not all Heidegger scholars agree with all of Dreyfus's interpretations.

[94] To avoid confusion, I cite the main text of *What is Metaphysics?* as Heidegger (1929), the postscript as Heidegger (1943), and the introduction as Heidegger (1949).

[95] Heidegger (1943), p. 277. Kaufmann notes (pp. 39 and 234) that his translation 'thinking that recalls' had Heidegger's 'enthusiastic approval'.

[96] A surreal, if not psychotic, depiction of a *failed* subject/object (or mind/matter) relation with the world is to be found in Jean-Paul Sartre's novel *Nausea*, which takes its title from the nauseous disgust that comes upon its narrator-diarist, Antoine Roquentin, as he progressively discovers the brute existence of things. One of Roquentin's discoveries is that things have 'broken free from their names' and, recalling the red plush ('thousands of little red paws in the air, all stiff, little dead paws') of the seats in a tram, he writes, 'They are there, grotesque, stubborn, gigantic, and it seems ridiculous to call them seats or say anything at all about them: I am in the midst of Things, which cannot be given names.' Sitting on a bench in the park, he is fascinated and horrified by the root of a chestnut tree, which he finds he cannot pin down by repeating '"It is a root".... [because] you could not pass from its function as a root, as a suction-pump, *to that*, to that hard, compact sea-lion skin, to that oily, horny, stubborn look.' Sartre (1938), pp. 180 and 186. Heidegger might say that Roquentin only experiences his world breaking down like this because he mistakenly believes that his proper relation to it is that of a subject among objects. See also my n. 163 below.

# Notes

[97] Heidegger (1943), p. 267.

[98] Heidegger (1927); p. 67 in the Macquarrie and Robinson translation, the source of all my quotations from *Being and Time*. I incorporate some of the modifications to that translation made in Dreyfus (1991), with two main exceptions: (1) as here, I retain the capital letter 'B' that Macquarrie and Robinson use for 'Being [*Sein*]' and related phrases; (2) as will come up later, I also retain their phrase 'the "they"' to translate Heidegger's '*das Man*'. Although Dreyfus has persuasive reasons (his pp. 11 and 151-52) for translating *Sein* as 'being' with a small 'b' and *das Man* as 'the one', I feel that Being and the 'they' are preferable, not least because lower-case 'being' for *Sein* invites confusion with 'a being' in the quite different sense of 'an entity'.

[99] Heidegger (1949), p. 272.

[100] I have purloined the phrase 'transparent coping' from Dreyfus, who is very good (better than the man himself, in my opinion) at conveying the import for Heidegger of the trivial, everyday experience of opening a door or hammering in a nail. See particularly pp. 257-61 in Dreyfus (1987) and, for the 'transparency' of 'transparent coping', the discussion of the blind person's cane in Dreyfus (1991), p. 65.

[101] And under such circumstances, according to Heidegger, we experience *un*-ready-to-hand door handles and hammers as akin to 'present-at-hand [*vorhanden*]' entities, that is, with a particular kind of Being that belongs to things other than Dasein. See also n. 168 below.

[102] Heidegger (1949), p. 271.

[103] Dreyfus records that Heidegger's reaction to the German translation of *Being and Nothingness* was to ask rhetorically, 'How can I even begin to read this muck [*Dreck*]?' Dreyfus himself is very sharp, and very funny, on Sartre's 'brilliant misunderstanding' of *Being and Time*: 'he read Heidegger and was converted to what he thought was Heideggerian existentialism. But as a Husserlian and a Frenchman he felt he had to fix up Heidegger and make him more Cartesian.' Dreyfus (1987), p. 275.

[104] I return to Sartre's consciousness-creation story on my pp. 86-87.

[105] Heidegger (1927), p. 89.

[106] Ibid., p. 67. Heidegger's italics.

[107] Ibid., p. 32.

[108] Ibid., p. 27.

[109] Ibid., p. 25. Heidegger's italics. Elsewhere (e.g., pp. 32-33, 168) he uses the phrase 'pre-ontological understanding' more or less interchangeably with 'average understanding'. I think I prefer the simpler wording of 'we already live in an understanding of Being [*wir je schon in einem Seinsverständnis leben*]': ibid., p. 23.

# Notes

[110] Ibid., pp. 28 and 32.

[111] I gather 'ownmost' was coined by Macquarrie and Robinson to translate Heidegger's adjective *'eigensten'*: ibid., p. 36.

[112] Ibid., p. 33.

[113] Ibid., p. 34. Here I am stripping away most of Heidegger's technical language in the hope of grasping his underlying argument more clearly. In Heidegger's own terms, Dasein's capacity to understand all modes of Being is the third of three 'priorities' and provides 'the ontico-ontological condition for the possibility of any ontologies.' You see what I mean.

[114] My hijacking is, of course, from the opening lines of Epistle II of Pope's *Essay on Man*: 'Know then thyself, presume not God to scan, / The proper study of mankind is man.'

[115] I am fast discovering that one of the difficulties of writing about Heidegger is deciding how to arrange the material, and what to omit. My arrangement of material is fairly arbitrary, and I omit a great deal (most obviously the whole theme of temporality, the *Time* of *Being and Time*), but I hope that what follows brings out at least some of the striking originality of Heidegger's early philosophy.

[116] Heidegger (1927), pp. 79-80. According to Macquarrie and Robinson (p. 80 footnotes), Heidegger's etymological claims in 'this puzzling passage' were heavily influenced by articles on archaic German by Jacob Grimm.

[117] I allude to the third of Heidegger's four 'significations' of the word 'world', which defines it as 'that "wherein" a factical Dasein ... can be said to "live"': ibid., p. 93. Also the commentary in Dreyfus (1991), pp. 89-91. For 'factical [*faktisch*]', akin in nuance to the noun 'facticity [*Faktizität*]', see my n. 152 below.

[118] Dreyfus suggests 'inhabiting' as a way to describe Being-in's mode of Being. 'When we inhabit something, it is no longer an object for us but becomes part of us and pervades our relation to other objects [sic] in the world': Dreyfus (1991), p. 45.

[119] Heidegger (1927), p. 84. Heidegger's italics.

[120] Dreyfus points out that for Heidegger entities or beings encounter Dasein, rather than the other way round, and that generally it would be better to translate his word *'begegnen'* as 'things *show up* for [Dasein]': Dreyfus (1991), Preface p. x

[121] Heidegger (1927), p. 97. In Heidegger's terms , 'Equipment is essentially "something in-order-to ... [*etwas, um zu* ...]".' 'Taken strictly,' he stresses, 'there "is" no such thing as *an* equipment.'

[122] Ibid., p. 116.

[123] Ibid., p. 120.

# Notes

[124] Here I am closely following Dreyfus, who observes that 'Heidegger ... relates Dasein's openness to the tradition, which extends from Plato to the Enlightenment, of equating intelligibility with illumination': Dreyfus (1991), p. 163.

[125] Heidegger (1927), p. 171. Heidegger's italics.

[126] Ibid., p. 171. Heidegger's italics.

[127] Ibid., p. 149. Although Heidegger's term 'everydayness [*Alltäglichkeit*]' is of key importance, as we are about to see, he does not get round to defining it himself until p. 422: '"Everydayness" manifestly stands for that way of existing in which Dasein maintains itself "every day".... But what we have primarily in mind ... is a definite "*how*" of existence by which Dasein is dominated through and through "for life".'

[128] Ibid., p. 167. Heidegger's italics.

[129] Ibid., p. 154-55. Heidegger's italics. He elaborates on the idea that Dasein's self derives from — is 'given' by — the 'they' in a related passage on p. 167: '*Primarily*, it is not "I", in the sense of my own self, that "am", but rather the Others, whose way is that of the "they". In terms of the "they", and as the "they", I am primarily "given" to "myself". Primarily Dasein is "they", and for the most part it remains so.'

[130] Ibid., p. 163.

[131] To avoid this sort of misperception, Dreyfus suggests that we think of human beings as 'hav[ing] Dasein in them': Dreyfus (1991), p. 95.

[132] Heidegger (1927), p. 164.

[133] Ibid., p. 164. Heidegger's italics.

[134] Ibid., pp. 164-65. The 'dictatorship' and 'publicness' are all the more insidious, Heidegger suggests, for asserting themselves in those attitudes and behaviours that we most take for granted: 'We take pleasure and enjoy ourselves as *they* take pleasure; we read, see, and judge about literature and art as *they* see and judge; likewise we shrink back from the "crowd" as *they* shrink back; we find "shocking" what *they* find shocking.'

[135] Ibid., p. 155. Heidegger's italics. No prizes for guessing that he does not mean 'concerned' in its straightforward dictionary sense. Usually translated by the noun 'concern', in Heidegger the verb '*besorgen*' refers to Dasein's transparent coping in relation to ready-to-hand equipment (i.e., inanimate things). He has a different word, 'solicitude [*Fürsorge*]', for Dasein's transparent coping in relation to other Daseins (other human beings).

[136] Ibid., p. 283. There is a nice symmetry here with Heidegger's comment (p. 163) that 'Others ... *are* what they do'.

[137] Ibid., p. 405. The phrase 'knows its way about' is the translation of '*sich ..."auskennt"*' chosen by Macquarrie and Robinson.

# Notes

[138] Ibid., p. 167. Averageness is one aspect of publicness. There are others, including 'levelling down [*Einebnung*]', which describes Dasein's socialized tendency to 'level' all possibilities of Being (i.e., avoid or suppress possibilities that might make it stand out as different) so that they conform with what the 'they' would consider right and proper (or 'valid'): ibid., p. 165.

[139] Ibid., p. 41.

[140] Ibid., p. 119.

[141] '*Worum-willen*' is one of the terms, and the most important, that Heidegger applies to the structure of involvements. For a succinct explanation, see Dreyfus (1991), pp. 91-96.

[142] Heidegger (1927), p. 116.

[143] Better late than never, I should explain that there is no special significance in the fact that I keep coming back to the carpenter's hammer. It is an example that Heidegger himself gives, but of course the same point could be illustrated with any piece of equipment and any for-the-sake-of-which associated with it, such as the telescope that an astronomer uses for the sake of being (or Being?) a professional astronomer.

[144] As evidence that Heidegger conceives of clearing as a shared situation, Dreyfus observes that the text (as in the passage quoted on my p. 43) speaks of '*the* clearing', not '*its* [i.e., Dasein's] clearing', and he offers an analogy to clarify the distinction: if a group of people are working together to make a clearing in the forest, each individual clears their own patch and yet the end result is a single clearing, '*the* clearing': Dreyfus (1991), pp. 164, 353 [n. 1].

[145] Heidegger (1927), pp. 401-02.

[146] Heidegger made this comment in a conversation with Dreyfus: Dreyfus (1991), p. 239.

[147] There is general agreement, however, that the Macquarrie and Robinson translation 'state-of-mind' is plain wrong because it implies a 'mental state', which is definitely not what Heidegger intends. Dreyfus, who explains how Heidegger coined '*Befindlichkeit*' from the everyday greeting '*Wie befinden Sie sich?*' [roughly equivalent to 'How are you doing?'], suggests 'affectedness' as a somewhat better translation: ibid., p. 168.

[148] Strictly speaking, as I understand, 'mood' is an example or subset of 'attunement'.

[149] Heidegger (1927), p. 176.

[150] Ibid., p. 175.

[151] Ibid., p. 174. Heidegger's italics.

[152] 'Thrownness' is a synonym for 'facticity [*Faktizität*]', a term I do not use in my main text but which likewise expresses the idea that Dasein *finds* itself

# Notes

'delivered over to entities ... it needs in order to be able to be as it is': ibid., p. 416. Heidegger's phrase '*not* of its own accord' occurs on p. 329.

[153] Ibid., p. 225.

[154] Ibid., p. 185. Elsewhere (p. 329) he equates projection with *existence* — another illustration, I think, of the chameleon nature of Heidegger's individual terms and yet, at the same time, of the broad consistency of his language as a whole.

[155] Ibid., p. 185.

[156] Ibid., p. 236. Heidegger's italics. His phrase '[e]ssentially ahead of itself' (p. 458) is surely the origin of Sartre's expression 'I await myself in the future [*je m'attends dans le futur*]'. Metaphorically speaking (perhaps literally?) and as noted earlier (my main text p. 40), a lot was lost in translation when Sartre read Heidegger, for Sartre's 'I' is the 'for-itself', a conscious subject in a world of objects, the exact opposite of a Dasein in Heidegger's sense: Sartre (1943), pp. 36 and 39.

[157] Or as Dreyfus sums it up more comprehensively: 'Heidegger wants to point out that in everyday transparent, skilled coping, Dasein is simply oriented toward the future, doing something now in order to be in a position to do something else later on.... Moreover, what it makes sense to do at any moment depends on the background of shared for-the-sake-of[-whichs] available in the culture.... Dasein is always already in a space of possibilities offered by the culture, and it normally presses forward into one of these possibilities without standing back and choosing what to do. All this Heidegger calls understanding.' Dreyfus (1987), p. 265.

[158] Heidegger (1927), p. 331.

[159] Ibid., p. 231. Heidegger's italics. His word 'obstinacy [*Aufsässigkeit*]' plainly has the nuance 'disturbing to Dasein'; elsewhere (p. 103) it has the more specific nuance 'disturbing to Dasein's usual mode of dealing with the world (i.e., transparent coping)'.

[160] Ibid., p. 230.

[161] Ibid., p. 235.

[162] Ibid., p. 233, including Macquarrie and Robinson footnote 1.

[163] Heidegger (1929), p. 51. For consistency, I have substituted 'in suspense' and 'suspendedness' for Krell's 'hover' and 'hovering'. Despite his similarly graphic imagery, Heidegger is *not* describing the sort of dissolution of the world experienced by Roquentin in Sartre's novel *Nausea*: my n. 96 above.

[164] Ibid., p. 53.

[165] Heidegger (1927), p. 330. Heidegger starts by saying that Dasein *is* 'the basis of its potentiality-for-Being' but then explains that, because it is a *thrown*

# Notes

basis, 'being a basis' (i.e., 'existing as thrown') means that Dasein itself '*is* a nullity of itself'.

[166] Ibid., pp. 233 and 232.

[167] Ibid., p. 393. The following two quotations in this paragraph are from the same page.

[168] As touched on in my n. 101 above, things that go wrong help to reveal 'presence-at-hand [*Vorhandenheit*]', one of the kinds of Being that belong to entities other than Dasein (i.e., non-human kinds of Being). As I understand Heidegger's 'existential conception of science' (his phrase), the stuff of scientific theory is, by definition, the stuff of presence-at-hand: that is, what scientists study are not pieces of *ready-to-hand* [*zuhanden*] equipment that only make sense in relation to some human for-the-sake-of-which, but *present-at-hand* [*vorhanden*] entities with properties that can be isolated, observed and theorized about 'objectively'. My primary source here is Heidegger (1927), pp. 408-15.

[169] Ibid., p. 33.

[170] Ibid., p. 275. Although I treat the 'undifferentiated' and 'inauthentic' as two distinct modes, Heidegger himself sometimes seems to regard them as synonymous. Dreyfus attributes this ambiguity largely to the intrusive influence of Kierkegaard: Dreyfus (1991), n. 58 on p. 361.

[171] Heidegger (1927), p. 232. Heidegger's italics.

[172] Ibid., p. 394.

[173] I hesitate to say 'capable of choosing authenticity' because, correctly speaking, Dasein does not choose but *accepts* 'the call' to authenticity that 'comes from the soundlessness of unsettledness': ibid., p. 343. For the ambiguous nature of Dasein's 'choosing', see Dreyfus (1991), pp. 316-19.

[174] Heidegger (1927), p. 354.

[175] Dreyfus observes that in Luther's translation of 1 Corinthians 15:51-52 '*Augenblick*' corresponds to the phrase 'in the twinkling of an eye' in the King James translation ('we shall all be changed, / In a moment, in the twinkling of an eye'). Dreyfus (1991), p. 321.

[176] Heidegger (1927), p. 224. My italics.

[177] Or 'takes a stand on', the phrase often used by Dreyfus in his commentary.

[178] Heidegger (1927), p. 346.

[179] Ibid., p. 167. Heidegger's account of the relation between *authentic* self and the *they-self* is confusing: well beyond the scope of my essay, the reasons for the confusion are discussed in Dreyfus (1991), pp. 239-41.

[180] Heidegger (1927), p. 347. Heidegger's italics.

[181] Ibid., p. 376.

# Notes

[182] Ibid., p. 395.

[183] 'Equanimity [*Gleichmut*]' is the word Heidegger himself uses to describe Dasein's authentic mood in the face of death (which, to cut a long story short, is another name for anxiety in Heidegger's vocabulary): ibid., p. 396.

[184] Ibid., p. 358.

[185] Sartre (1943), pp. 439 and 39. Sartre's absurd freedom is exemplified by the gambler, who walks away from the gaming table one day because he has 'freely and sincerely decided not to gamble any more' and yet may be back at the table the very next day, faced all over again with the recognition that '*nothing* prevents [him] from gambling.' Ibid., pp. 32-33

[186] Heidegger (1927), p. 225.

[187] An example of Heidegger's ambiguous use of the terms 'undifferentiated' and 'inauthentic': see my n. 170.

[188] The other two are thrownness and projection: pp. 49-50 in my main text.

[189] Heidegger (1927), pp. 219-20. Chapter 13 of Dreyfus (1991) is devoted to disentangling the complexities and inconsistencies of 'falling'.

[190] Ibid., p. 221. Heidegger's italics. I suspect Dreyfus would say that Heidegger is not alluding directly to the biblical Fall but to Kierkegaard's interpretation of it, another very complicated story beyond the scope of my essay: Dreyfus (1991), pp. 229, 315 and 335.

[191] Heidegger (1927), p. 233.

[192] Ibid., p. 312.

[193] Ibid., p. 222. The 'they' has several means of creating and perpetuating the illusion of meaningfulness, including a form of language that Heidegger calls 'idle talk [*Gerede*]'. They are all beyond my scope.

[194] Ibid., p. 422. While I believe 'has not chosen' is correct as a dictionary rendering of Heidegger's '*nicht ... gewählt hat*', it is misleading here for reasons touched on in my n. 173.

[195] Ibid., p. 68. Heidegger's italics.

[196] I have borrowed the word 'disowning' from Dreyfus, whose commentary I find particularly valuable here: Dreyfus (1991), pp. 26-27 and 315.

[197] Heidegger (1927), p. 69.

## 6. Nothingness and suchness

[198] This and the following verse passage are my own reworkings of John Bester's translations from the Chinese in Awakawa (1971), pp. 14-15. The Sanskrit word '*bodhi*' means 'awakening' or 'enlightenment'; the original

# Notes

*bodhi* tree was the sacred tree under which the historical Buddha, Gautama, is said to have achieved enlightenment.

[199] As I understand it, there is no reliable evidence that Huineng and Shenxiu were at the monastery at the same time, or even that Hongren designated Huineng as his sole successor. My primary source is John McRae's translation of *The Platform Sutra of the Sixth Patriarch* in McRrae (2000).

[200] The Sanskrit word '*śūnyatā*' comes from the adjective '*śūnya*', which means 'empty'— hence the noun 'emptiness' or, as I express it, 'nothingness'. Two Japanese terms used to translate '*śūnyatā*' are '*kū* [空]' meaning 'void' or 'emptiness', and '*mu* [無]' meaning 'nothing' or 'nothingness'. Among other synonyms, *śūnyatā* is sometimes referred to as 'formless form', as in Abe (1997). *Śūnyatā* should properly be discussed in relation to the Buddhist ontological principle of 'dependent origination [*pratītyasamutpāda* in Sanskrit; *engi* 縁起 in Japanese]': beyond my scope but I am grateful to John Krummel (private correspondence) for pointing this out.

[201] 'Suchness' is a translation of the Sanskrit word '*tathatā*'. My own phrase 'just as it is' is a literal rendering of the colloquial Japanese phrase '*sono mama* [そのまま]', which frequently occurs in everyday speech. In the formal language of Japanese Buddhism, the equivalent of *tathatā* is '*shinnyo* [真如]'. In the Chan/Zen illustrated text known as the *Ten Ox-Herding Pictures*, suchness is expressed in the Chinese text accompanying Picture Nine ('Returning to the Source', which simply shows a tree in bloom beside a river) as follows: 'Boundlessly flows the river, just as it flows. Red blooms the flower, just as it blooms.' This translation comes from the commentary on Picture Nine (and its relation to the empty circle depicted in Picture Eight, 'Ox and Herder Both Gone'), in Ueda (1982), p. 15. The best known version of the *Ten Ox-Herding Pictures* is by the twelfth-century Chan priest Kuoan.

[202] See my n. 39 for the source of this Bacon quotation (n. 38 for Luther).

[203] Although the Kyoto School developed out of Nishida's philosophy, it was not founded by him in any formal sense.

[204] The sorry episode is briefly recounted in Heisig (2001), chaps. 25 and 26. Heisig's own judgement (p. 99) is that 'if [Nishida] is to be faulted for anything, it is for the failure to realize that ignorance of his own limitations was a kind of complicity.' See also the essays in Heisig and Maraldo (1995).

[205] In his final essay in 1945 Nishida went so far as to say that he thought his 'logic of *basho*' (of which more anon in my main text) could 'embrace' Kant's philosophy: Nishida (1945), Pt. I, p. 13; 西田 (1945)、317頁、「カント哲学を私の場所的論理の中に包容し得ると思う」。In the introduction to his translation David Dilworth invites us to think of Nishida's essay as 'the missing "fourth Critique"' to complement the three that Kant did write, *The Critique of Pure Reason*, *The Critique of Practical Reason*, and *The Critique of Judgement*: Dilworth (1987), p. 14. For the text of Nishida's essay itself I cite the translation by Yusa Michiko. For Nishida on Kant, see also my n. 266.

# Notes

[206] An impression evidently shared by native speakers of Japanese, even among Nishida's students and successors. Ueda Shizuteru, a leading figure in the Kyoto School and an authority on Nishida, is on record as saying that 'it is not unusual to read for pages with almost no idea of what it is that Nishida is trying to say': Ueda (1995), p. 175.

[207] Nishida (1926), p. 81; 西田 (1926)、118頁、「個々の音は音調に於てあるのである。」

[208] Nishida (1911), pp. 32-33; 西田 (1911)、59-60頁、「芸術家の精巧なる一刀一筆は全体の真意を現わす」。For illustrations of the kind of ink paintings Nishida may well have in mind, see Awakawa (1971), which contains a fine selection of 'zenga [Zen paintings]'; they include some by the Japanese artist Sesshū (1420-1506), who studied in China and had a lasting influence on Japanese landscape painting after his return.

[209] Not, it is worth noting, as *An Inquiry into Zen*: although pronounced the same in Japanese, the character [禅] for the Zen of Zen Buddhism is quite different from the character [善] meaning 'the good'.

[210] Nishida (1911), p. 3; 西田 (1911)、17頁、「経験するというのは事実其儘に知るの意である……純粋というのは、普通に経験といって居る者もその実は何らかの思想を交えて居るから、毫も思慮分別を加えない、真に経験其儘の状態をいうのである。」

[211] See my nn. 236 and 237 below.

[212] My own translation from 西田 (1945)、385頁、「最も遠くしてしかも最も近きものが、最も真なるものであるのである。何処まで行っても、その出立点を失わない、逆にこれに返るという立場において、真理が成立するのである。私の行為的直観というのは、これにほかならない。」Although only one grammatical subject ('truth [*shinri*]') is specified in Nishida's second sentence, I am persuaded by Yusa that another, human subject ('we' in my translation) has to be understood here because, according to Nishida's notion of 'acting-intuition', we *participate in* the 'arising' of truth by being in and acting upon the world. Yusa's translation of the same passage can be found in Nishida (1945), Pt. II, p. 111. See also my n. 213 following.

[213] Roughly speaking, 'acting-intuition' (also translated as 'active intuition') addresses the apparent contradiction that we are passively determined by the world we inhabit and yet actively act upon and shape it. For a brief account, see John Krummel's introduction to Nishida (1926), pp. 32-33. Heisig (2001), pp. 53-56, covers much the same ground.

[214] Nishida's concept of 'pure experience [*junsui keiken*]' was heavily influenced by — but, he would insist, was qualitatively very different from — the psychology of William James, Wilhelm Wundt, and others.

[215] Nishida (1911), Preface [1936] p. xxxiii; 西田 (1911)、10頁、「実在は現実そのままのものでなければならない」。

[216] Nishida (1911), p. 4; 西田 (1911)、17頁、「知識とその対象」。

# Notes

[217] Nishida (1911), p. 130;　西田 (1911)、198頁、「この意識の統一力なる者は決して意識の内容を離れて存するのではない、かえって意識内容はこの力に由って成立するものである。」

[218] My own translation of '*tōitsuteki arumono*'. For reasons explained in their footnote 7, Abe and Ives render the same phrase as 'a certain unifying reality': Nishida (1911), p. 7;　西田 (1911)、23頁、「統一的或る者」。

[219] Nishida (1911), p. 56;　西田 (1911)、92頁、「我々の思惟意志の根柢における統一力と宇宙現象の根柢における統一力とは直に同一である」。

[220] Adapted from the Abe/Ives translation in Nishida (1911), p. 62;　西田 (1911)、102頁、「苟も我々の知り得る、理会し得る世界は我々の意識と同一の統一力の下に立たねばならぬ。」

[221] Nishida (1911), p. 130;　西田 (1911)、199頁、「極めて主観的なる種々の希望の如き者」。

[222] Adapted from the Abe/Ives translation in Nishida (1911), pp. 130-31;　西田 (1911)、199頁、「真の意識統一というのは我我を知らずして自然に現われ来る純一無雑の作用であって、知情意の分別なく主格の隔離なく独立自全なる意識本来の状態である。我々の真人格は此の如き時にその全体を現わすのである。」The phrase 'pure and simple' alludes to a passage in the *Linji lu* (known as the *Rinzairoku* in Japanese), a collection of sayings attributed to Linji (Rinzai in Japanese), a revered Chan master who died in 866: Discourse XVIII in Sasaki (2009), pp. 26 and 257. See also p. 82 of my main text.

[223] This speculation aside, elsewhere in *An Inquiry into the Good* Nishida does explicitly liken our experience of the world of dualities to the biblical Fall. 'The Fall of Man,' he says, '[occurred] not only in the distant past of Adam and Eve, but occurs moment by moment in our own minds': adapted from the Abe/Ives translation in Nishida (1911), p. 170;　西田 (1911)、253頁、「人祖堕落はアダム、エヴの昔ばかりではなく、我らの心の中に時々刻々行われて居るのである。」

[224] '*Jikaku ni okeru chokkan to hansei*' [not included in my Sources], originally translated as 'Intuition and Reflection in Self-Consciousness'. For reasons explained in the endnotes of Heisig (2001), p. 293, the title now tends to be translated as 'Intuition and Reflection in Self-Awareness'. By the end of the work Nishida has come to the position that 'self-awareness' is synonymous with 'absolute will [*zettai ishi*]', that is, as Heisig puts it (p. 49), 'an absolute free will that subsumes within itself not only individual wills but the whole of reality.'

[225] From Nishida's preface to '*Hataraku mono kara miru mono e*' ['From the Acting to the Seeing': not included in my English-language Sources], here adapted from Krummel's quote in his introduction to Nishida (1926), p. 11;　西田 (1927)、36頁、「有るもの働くもののすべてを、自ら無にして自己の中に自己を映すものの影と見るのである、すべてのものの根柢に見るものなくして見るものという如きものを考えたいと思うのである。」

# Notes

[226] While I choose to focus on '*basho*' principally for reasons of content, there is also the practical constraint that it is beyond the scope of this essay (and my qualifications for the job) to examine every stage in the later development of Nishida's thinking. For a relatively brief, if dense, outline of Nishida's philosophy as a whole, see Krummel's introduction in Nishida (1926), pp. 3-48. Supplemented by copious notes and suggestions for further reading, it introduces important concepts and terms that I do not mention at all, including the tongue-twisting 'absolutely contradictory self-identity [*zettai mujunteki jikodōitsu*]'. Robert Carter's full-length study (seemingly based on an exhaustive review of the English-language literature) is a very readable account of Nishida's philosophy, thematically arranged: Carter (1997).

[227] Nishida (1911), p. xxxi; 西田 (1911)、9頁、「今日から見れば、この書の立場は意識の立場であり、心理主義的とも考えられるであろう。」

[228] I do not wish to overdo the solipsism question because, in his preface to the first edition, Nishida had congratulated himself on *avoiding* solipsism. 'Over time,' he wrote, 'I came to realize that it is not that experience exists because there is an individual, but that an individual exists because there is experience. I thus arrived at the idea that experience is more fundamental than individual differences, and in this way I was able to avoid solipsism': Nishida (1911), p. xxx; 西田 (1911)、6頁、「そのうち、個人あって経験あるにあらず、経験あって個人あるのである、個人的区別よりも経験が根本的であるという考から独我論を脱することができ」。

[229] Nishida (1911), p. 32; 西田 (1911)、58頁、「神秘的直覚」。

[230] Nishida (1928), p. 135; 西田 (1928)、249頁、「神秘的直観の世界」。

[231] Abe and Ives translate the same phrase as 'a certain mystical reality': Nishida (1911), p. 33; 西田 (1911)、61頁、「神秘的或る者」。Relates to my n. 218 above.

[232] Nishida (1945), Pt. II, p. 109; 西田 (1945)、382頁、「私の哲学を神秘的と考える人は、対象論理の立場から考える故である。」

[233] Adapted from the Yusa translation in Nishida (1945), Pt. II, p. 106; 西田 (1945)、378頁、「禅についての誤解は、すべて対象論理的思惟に基くのである」。Nishida applies the same critique to '[w]hat Western philosophy since Plotinus has called "mysticism"': even the One of Plotinus himself, Nishida insists, fails to get beyond 'the standpoint of object logic' [i.e., to speak of 'the One' implies a duality of 'One' and 'not-One']. Apropos Plotinus, see pp. 6-7 of my main text.

[234] Or, a direct translation of the word he coined, 'logicize' his philosophy: Nishida (1911), Preface [1936] p. xxxii; 西田 (1911)、10頁、「私の考を論理化する」。

[235] Ibid., p. xxxii; 同書、9-10頁、「一転して「場所」の考に至った。」

[236] The word chosen by, among others, Yusa in her translation of Nishida (1945). See also n. 237 following.

# Notes

[237] Nishida (1926), p. 50; 西田 (1926)、68頁、「イデヤを受取るものともいうべきものを、プラトンのティマイオスの語に倣うて場所と名づけて置く。」The key passages in the *Timaeus* are 49a, 50b-51a and 52a-b; pp. 67, 69-71 in Lee (1977). Krummel compares Plato's *chōra* with Aristotle's *topos* and suggests that, while *chōra* is 'certainly more significant' in relation to *basho*, *topos* may also have influenced Nishida's concept: Nishida (1926), p. 21 and n. 5 on pp. 188-89. Even *chōra* is far removed from *basho* because that has a philosophical ('logical') value in its own right, whereas Plato assigns *chōra* only a semi-mechanical function.

[238] My own translation from 西田 (1926)、67頁、「有るものは何かに於てなければならぬ」。This is one of many passages in the Krummel/Nagatomo translation where they use their newly-coined word 'implaced' (which has a passive form only) rather than 'placed': Nishida (1926), pp. 49, 188 [n. 2].

[239] Adapted from the Krummel/Nagatomo translation in Nishida (1926), p. 86; 西田 (1926)、126頁、「一般的なるものと一般的なるものと、場所と場所とが無限に重り合っているのである、限なく円が円に於てあるのである。」

[240] A literal translation of 「包論理的」 in 西田(1926)、72頁。The first character [包], pronounced '*hō*' in compounds, here has the connotation 'embrace' or 'envelop'. As a verb on its own, pronounced '*tsutsumu*', the same character has the everyday meaning 'to wrap', as in to wrap a parcel.

[241] Krummel and Nagatomo create the word 'peri-logical' to translate '*hōronriteki*': Nishida (1926), pp. 6 and 191 (note 25). 'Enveloping logic' seems adequate for my purposes.

[242] This interpretation of the gradations as degrees of reality is my own. It is more usual to speak in terms of relative 'concreteness' (the outermost circle most 'concrete', the innermost least 'concrete'), language I prefer to avoid because it risks sending us down the rabbit hole of Nishida's borrowings from Western philosophy (in this case, I believe, Hegel's concept of the concrete universal).

[243] Carter describes, with schematic diagrams, nine concentric circles in all, surrounded by what he labels 'the final enveloping field (*basho*) about which nothing can be said': Carter (1997), pp. 33-44. He mostly discusses *bashos* as 'universals' and has various names for the circles within circles, including 'stages', 'levels' and 'layers'. How about 'rings' or, by analogy with Ptolemaic cosmology, 'spheres' of *basho*?

[244] *Timaeus* 28-29b and 52a-c; pp. 40-41 and 71-72 in Lee (1977).

[245] Nishida (1911), Preface [1936] p. xxxii; 西田 (1911)、9頁、「ギリシャ哲学を介し」。

[246] Aristotle, *Metaphysics* 1029a; p. 711 in the translation in Cohen, Curd and Reeve (2000).

[247] Nishida defines the 'predicate-plane [*jutsugomen*]', or 'plane of consciousness [*ishikimen*]', as that which has 'enveloped the subject [i.e.,

# Notes

Aristotle's grammatical subject] of judgement': Nishida (1926), p. 96; 西田 (1926)、142頁、「意識面というのは判断の主語を包み込んだ述語面」。

[248] I am misappropriating rather than borrowing Heidegger's term, of course, because fundamentally there is no relation between his concept of the clearing (my main text, pp. 43-44) and Nishida's *basho* of consciousness.

[249] Nishida (1926), pp. 95, 213 [n. 268]; 西田 (1926)、141頁、「我とは主語的統一ではなくして、述語的統一でなければならぬ、一つの点ではなくして一つの円でなければならぬ、物ではなく場所でなければならぬ。」 The term 'X', which Krummel and Nagatomo have added to their translation, is not expressed in the Japanese and has to be understood from the context: 「すべての経験的知識には「私に意識せられる」ということが伴わねばならぬ」。 Nishida's phrase for 'predicating unity' is '*jutsugoteki tōitsu*'.

[250] The '*tairitsuteki mu no basho*'.

[251] Yet another name for the *basho* of consciousness is 'the *basho* of relative nothing [*sōtai mu no basho*]', which, like the *basho* of oppositional nothing, connotes that consciousness is only a nothing in relation to the beings that are other than itself.

[252] I reproduce here, more or less verbatim, Krummel's concise explanation of this particularly difficult aspect of the *basho* concept: Nishida (1926), p. 23 and diagram on p. 27. See also my n. 273 below.

[253] Nishida (1926), pp. 100-01; 西田 (1926)、148-49頁、「真の直観はいわゆる意識の場所を破って直にかかる場所に於いてあるのである。」 Nishida's terms for 'true intuition' and 'the *basho* of intuition' are '*shin no chokkan*' and '*chokkan no basho*' respectively.

[254] Quoted more fully on my p. 72. It is natural to assume that Nishida's notion of reality as a self-mirroring intuition is influenced by Buddhism, and surely it is (my n. 256 below). By Nishida's own account, however, the notion was first suggested by his reading of the American philosopher Josiah Royce, who imagined a self-representative system as a map that endlessly maps itself mapping itself. In the original preface to his 1917 work (title in my n. 224), Nishida refers explicitly to Royce's 'Supplementary Essay' to his Gifford Lectures of 1899, in Vol. 1 of *The World and the Individual*.

[255] I would like to think that the image of a ceaseless flow of reality may even evoke Heraclitus's river into which no one can step twice, symbolizing, in Nishida's words, that 'Reality is a succession of events that flow without stopping': Nishida (1911), p. 54; 西田 (1911)、89頁、「実在は流転して暫くも留まることなき出来事の連続である。」

[256] Adapted from the Krummel/Nagatomo translation in Nishida (1926), p. 61; 西田 (1926)、86頁、「有が真の無に於てある時、後者が前者を映すというのほかはない。映すということは物の形を歪めないで、そのままに成り立たしめることである、そのままに受け入れることである。映すものは物を内に成り立たしめるが、これに対して働くものではない。」 Krummel suggests that Nishida's imagery of mirroring and self-mirroring is comparable

# Notes

with the 'great mirror wisdom' of Yogācāra Buddhism: Nishida (1926), n. 9 on p. 189 and nn. 85, 87 and 88 on p. 197.

[257] That is, it makes room for them, just as, at its own level, the *basho* of consciousness makes room for individual beings by enveloping them: pp. 76-77 of my main text.

[258] Adapted from the Krummel/Nagatomo translation in Nishida (1926), p. 57; 西田 (1926)、80頁、「真の無の場所というのは如何なる意味においての有無の対立をも超越してこれを内に成立せしめるものでなければならぬ。」

[259] Nishida (1911), p. xxx; 西田 (1911)、6頁、「哲学の終結」。

[260] Nishida (1928), p. 133; 西田 (1928)、247頁、「絶対に自己を否定して、見るものなくして見、聞くものもなく聞くものに至るのが宗教的理想である」。

[261] Suggested by Krummel in his introduction to Nishida (1926), p. 26. He offers 'existential' in the context of what he calls the 'specifically human significance of "life-and-death" … awareness of the very finitude of one's being vis-à-vis death'.

[262] Nishida (1911), pp. 79-80; 西田 (1911)、129頁、「此の如き神の考は甚だ幼稚であって」。'God [*kami*],' Nishida insists, 'is not something that transcends reality…. [but] a fundamental spiritual principle at the base of reality…. God is the great spirit of the universe.' Writing in 1911, Nishida claims that this idea of God 'accords with the fundamental truth of Indian religion'; on the evidence of Nishida's later writings, I think we can say that it also accords with (or developed into) his own conception of reality as a self-determining, self-mirroring whole — in other words, for Nishida 'God' came to be yet another synonym for the *basho* of true nothing.

[263] Nishida (1911), p. 149; 西田 (1911)、223頁、「真正の宗教は自己の変換、生命の革新を求めるのである。」

[264] From Abe's introduction to Nishida (1911), p. xix.

[265] Nishida (1911), p. 126; 西田 (1911)、192頁、「自己の真実在と一致するのが最上の善ということになる。そこで道徳の法則は実在の法則の中に含まるる様になり、善とは自己の実在の真性より説明ができることとなる。」

[266] 'Reason with a capital "R"' is my phrase, not Nishida's. What I mean by it is the Enlightenment faith in a Universal Reason, the laws of which, once discovered, would enable mankind to remake the world and achieve true knowledge, virtue, justice and happiness. In my essay 'The Frog and the Basilisk' I suggest that one way of understanding the misplaced faith in Universal Reason is to see it as a secularized manifestation of the Word of God: Wilkinson (2015), pp. 47-48. Nishida himself speaks of 'mere reason' (or '*blosse Vernunft*' in German), as when he reproaches Kant for regarding religion as no more than a 'supplement' to morality ('moral reason'), whereas his own view is that religion does not fit 'within the limits of "mere reason"':

# Notes

Nishida (1945), Pt. I, pp. 2-3; 西田 (1945)、301 頁、「「単なる理性」 *blosse Vernunft* の中には、宗教は入って来ないのである。」See also my n. 205 above.

[267] Nishida discusses Epicureanism and utilitarianism in his chapter on 'hedonistic theory': Nishida (1911), pp. 115-21; 西田 (1911)、177-86 頁。

[268] Ibid., p. 125; 同書、191 頁、「善とは自己の発展完成 self-realization であるということができる。」The word 'self-realization', which Nishida writes in English, is believed to be an allusion to the nineteenth-century British philosopher Thomas Hill Green, whose *Prolegomena to Ethics* Nishida had studied early in his career.

[269] Ibid., p. 125; 同書、191 頁、「竹は竹、松は松と各自その天賦を充分に発揮するように、人間が人間の天性自然を発揮するのが人間の善である。」

[270] 'Go to the pine to learn about the pine. Go to the bamboo to learn about the bamboo. Set aside all personal thoughts and motives, for you will learn nothing if you insist on interpreting objects as *you* see them.' The saying, recorded in the *Akazōshi* ('Red Notebook', part of a collection of Bashō's teachings compiled by one of his disciples, Dohō, in the first years of the eighteenth century), is explored at some length in a collection of my own essays: Wilkinson (2019).

[271] The word '*byōjōtei*' has no equivalent in English. Yusa translates one of Nishida's phrases containing the word as 'an extremely ordinary standpoint' and then glosses it '[i.e., the horizon of everyday existence]', which is probably as close to a definition of *byōjōtei* as it is possible to get: Nishida (1945), Pt. II, p. 106; 西田 (1945)、378 頁、「極めて平常底なる立場」。

[272] Adapted from Ruth Sasaki's translation of Discourse XII in the *Linji lu*: Sasaki (2009), pp. 11-12 and 185. See also my n. 222 above.

[273] Adapted from the Abe/Ives translation in Nishida (1911), p. 143; 西田 (1911)、216-17 頁、「自己と宇宙とは同一の根柢をもって居る、否直に同一物である……実在の真善美は直に自己の真善美でなければならぬ。」In terms of Nishida's concept of *basho*, truth, good and beauty are the three key values or ideals (reflections of the *basho* of true nothingness) that guide the will in determining acts of consciousness: see p. 77 of my main text. In a reference to the neo-Kantian philosopher Wilhelm Windelband, Nishida describes truth, good and beauty as 'these [three forms of] consciousness of value' and insists that religious value can only be found in 'the fundamental relation' between them: Nishida (1928), p. 133; 西田 (1928)、247 頁、「宗教的価値は唯これらの価値意識に共通なる根本的関係……に求めねばならないとなし」。

[274] For an outline of the Buddhist doctrines and texts that may have had the greatest influence on Nishida, see the early chapters of Waldenfels (1980).

# Notes

## 7. Conclusions

[275] Nagel (2012), p. 29; quoted more fully on my p. 5.

[276] Ibid., p. 20.

[277] Ibid., p. 67.

[278] Ibid., p. 85.

[279] Ibid., p. 86.

[280] Ibid., p. 17.

[281] The references that follow are to the text of Wheeler's presentation, 'Beyond the Black Hole', collected in the volume *Some Strangeness in the Proportion*. Cited hereafter as Wheeler (1979).

[282] Wheeler (1979), pp. 362 and 358.

[283] I am deliberately omitting most of the technical detail. Conceptually, the crucial point is that the same photons behave differently and create different patterns (of intensity variation or interference), depending on how the experiment has been set up: it is as if the photons 'know' how the apparatus is configured, and that has been determined by the experimenter. For a more detailed but very accessible account of the standard double-slit experiment and Wheeler's modification of it, personally I recommend Gribbin (1992), pp. 164-71 and 210-11.

[284] The modification that Wheeler envisaged (adding special lenses to the standard setup) was not technically possible when he conceived of his thought-experiment, but equivalent modifications have since been made and demonstrated under laboratory conditions. The first major demonstration, I understand, was achieved in 2007 by a team led by Vincent Jacques of the Laboratoire de Photonique Quantique et Moléculaire in Cachan, France: technical report in Jacques et al. (2007). Needless to say, I am not suggesting that proof of the technical feasibility of Wheeler's delayed-choice 'mechanism' in the laboratory is the same as proof of its applicability to the universe as a whole.

[285] Wheeler (1979), pp. 358-59.

[286] 'The being of consciousness qua consciousness is to exist *at a distance from itself* [*à distance de soi*] as a presence to itself': Sartre (1943), p. 78. Sartre's italics.

[287] Ibid., p. xli [Sartre's Introduction].

[288] Ibid., p. xlii.

[289] Ibid. p. 84.

[290] Ibid., p. 81.

[291] Ibid., p. 620.

# Notes

[292] Ibid., p. 79.

[293] *On the Nature of the Universe*, Book 2.1058-61; p. 64 in Latham (2005). Quoted earlier on my p. 22.

[294] Ibid., Book 5.435-36; p. 139 in Latham. My p. 20.

[295] Ibid., Book 5.417-25; p 139 in Latham.

[296] Apropos this question, see my n. 262 above.

[297] The expressions 'somethingness' and '[not] completely shaken free from the character of being' are Abe's rather than Nishida's: Abe (1988), pp. 366 and 368 respectively.

[298] Or as he put it even before had developed the concept of *basho*, 'The fundamental mode of reality is such that reality is one while it is many and many while it is one': Nishida (1911), p. 57; 西田 (1911)、93頁、「実在の根本的方式は一なると共に多、多なると共に一」。

[299] Abe suggests that this can be expressed in terms of formal logic. He argues that in a 'truly subsumptive judgement' (which gives rise to what Nishida calls a 'concrete universal') a particular entity 'preserves all of its particularity, which is enveloped within something more universal. It does not lose its specific difference as it does when subsumed by an abstract universal — instead, [the] particular is grasped as the self-determination of a concrete universal…. A universal in this sense differs from an abstract universal, for it includes the principles of particularization and individualization': Abe (1988), p. 360. I understand the distinction between 'concrete' and 'abstract' universals can be traced back to Hegel: my n. 242.

[300] See my n. 256 for the source of this quotation.

[301] Nishida (1911), p. 94; 西田 (1911)、145頁、「意志は我々の意識の最も深き統一力であって、また実在統一力の最も深遠なる発現である……いわゆる自然は意志の発現であって、我々は自己の意志を通して幽玄なる自然の真意義を捕捉することができるのである。」

[302] Ibid., p. 81; 同書、131頁、「無限なる実在の統一力が潜んで居る、我々はこの力を有するが故に学問において宇宙の真理を探ることができ」。

[303] Ibid., pp. 80-81; 同書、130頁、「世界が或る目的に従うて都合よく組織せられてあるという事実から、全智なる支配者がなければならぬと推理するには、事実上宇宙の万物が尽く合目的に出来て居るということを証明せねばならぬ、しかし、これは頗る難事である。」

[304] See my nn. 163 and 164 for the sources of this passage and the following phrase 'held out into'.

[305] Source in my n. 134.

[306] *On the Nature of the Universe*, Book 5.417-22; p 139 in Latham. My p. 20.

# Notes

[307] I do not mean to imply that Bohr had no doubts or hesitations, or that he always expressed himself consistently. See, for example, my nn. 85 and 86.

[308] See my n. 87 for the source of this quotation.

[309] From Leon Rosenfeld's recollection of a conversation with Bohr: Rosenfeld (1963), p. 54. The conversation probably happened in Copenhagen but we are not told when, and nor is it clear whether the striking phrase 'better than any religion' is Bohr's own or Rosenfeld's.

[310] Bohr (1962), p. 4/12.

[311] Ibid., p. 8/12. I hope that I have interpreted this passage correctly, although I find Bohr's English very disjointed and I am assuming that among physicists the paper Einstein published in 1917 (proposing a quantum theory of radiation) is at least as important as the paper of 1905 in which he proposed the quantization of electromagnetic energy as an explanation for the photoelectric effect (my main text p. 11).

[312] The symposium, held in Trieste in September 1972 to celebrate the seventieth birthday of Paul Dirac, was attended by, among others, Dirac himself, Heisenberg and Wheeler. The proceedings are collected in the volume *The Physicist's Conception of Nature*, edited by Jagdish Mehra.

[313] d'Espagnat (1972), p. 734. My italics.

[314] Ibid., p. 730. Or as d'Espagnat puts it elsewhere (pp. 734-35), 'elementary scientific teaching implicitly propagates the view (A) that a kind of naïve realism is true; (B) that the ultimate reality is essentially constituted of an immense number of small elements, each possessing a fixed number of definite properties; and (C) that the local and causal interactions of these elements lead to combinations that account for the complexity of the actual world.'

[315] Ibid., p. 734.

[316] Ibid., p. 735. His italics.

[317] My source is Stadermann et al. (2019), especially Tables V and VI. In Belgium (Flemish-language curriculum) teachers have the option of including complementarity in their courses but are not obliged to; in the other 12 European countries (Austria, Denmark, Finland, France, Germany, Italy, the Netherlands, Norway, Portugal, Spain, Sweden and the United Kingdom) it is a required element. Findings for Canada relate only to Ontario. Although the study is limited in geographical scope, its intention is not, as the authors caution (p. 5), 'to give a complete overview of all countries around the world in which [quantum physics] is taught in secondary schools' but 'to analyze which content is typically used to introduce this challenging topic in different educational systems.' For comparison, the most recent (2015) survey of physics courses at upper secondary level in the series Trends in International Mathematics and Science Study (TIMSS Advanced 2015: cited in reference [20] in Stadermann et al.) includes brief information on elements of quantum physics in the curricula of nine European and non-European countries (France,

# Notes

Italy, Lebanon, Norway, Portugal, the Russian Federation, Slovenia, Sweden, and the United States).

[318] And possibly since the 1950s. Because of the fragmented nature of the German education system, we may never know exactly what was taught where and when, especially prior to reunification in 1990. My thanks to Rainer Müller of Technische Universität Braunschweig for doing his best to clarify.

[319] In Saxony and Lower Saxony (also in Norway), quantum entanglement (the 'action at a distance' that so troubled Einstein in connection with the EPR thought-experiment: my pp. 32-34) is already included in physics courses at upper secondary level. Another candidate for inclusion in general physics courses elsewhere?

[320] The everyday macroscopic model does *not* work, of course, for the purposes of particle physicists, electronics engineers, cryptographers and others professionally involved in quantum theory and quantum technologies. Having said that, I find it disappointing that the typical, if not dominant, position among them seems to remain instrumentalism, i.e., they treat quantum mechanics as a device of great theoretical and practical value but do not engage with its philosophical implications. See also my n. 88 ('Shut up and calculate').

[321] Bohr (1962), p. 4/12. Grim as it is, Bohr means his allusion to Bruno and Galileo to encourage optimism: over time, he suggests, even a system of thought that has been misunderstood and denounced as an outrage may quietly gain acceptance. The system of thought Bohr has in mind is complementarity, but I would like to think that the same principle applies to a system of thought, which I am calling '"natural" knowledge', that explicitly recognizes the unintelligibility of the world.

[322] 'And new philosophy calls all in doubt, / The element of fire is quite put out, / The sun is lost, and th' earth, and no man's wit / Can well direct him where to look for it. / And freely men confess that this world's spent, / When in the planets and the firmament / They seek so many new; they see that this / Is crumbled out again to his atomies. / 'Tis all in pieces, all coherence gone'. From Donne's *The First Anniversary: An Anatomy of the World* (1611), ll. 205-13.

[323] For the non-spatial sense of Heidegger's preposition 'in', see my p. 42. For a brief discussion of thrownness in relation to Dasein's receptiveness or 'moods', my pp. 48-49.

[324] Heidegger (1927), p. 171. Quoted more in context on my p. 44.

[325] From Heidegger (1929), p. 49. Quoted more fully on my p. 37.

[326] My interpretation of Heidegger's third 'signification' of 'world': Heidegger (1927), p. 93, and my n. 117. On the significant detail that Dasein is encountered by beings and entities rather than encounters them, my n. 120.

[327] Heidegger (1927), p. 120. See also my p. 43.

# Notes

[328] I stress again that this description does not apply to those special circumstances in which we experience things as 'present-at-hand', i.e., as having properties that can be isolated and observed in subject/object terms or 'objectively' — the relation to the world that Heidegger regards as secondary or derivative. See also my nn. 101 and 168.

[329] '[A]uthentic existence is not something which floats above ... everydayness; existentially, it is only a modified way in which ... everydayness is seized upon': Heidegger (1927), p. 224; previously quoted on p. 56 of my main text.

[330] A separate but related issue, I think, is that Heidegger's complex lexicon and abstract theorizing are not well suited to conveying the actual consequences of 'living' inauthentically (or in the 'undifferentiated' mode: my p. 54). He speaks of 'the tranquillized self-assurance' of Dasein, for example (my p. 59), but does not spell out what that means in simpler human terms. It is ironic that Sartre, who misunderstood Heidegger and turned Dasein into the 'for-itself' (my p. 40), does a much better job of depicting what he calls 'bad faith' than Heidegger does depicting 'tranquillization' or other aspects of Dasein's inauthenticity. The for-itself's tendency to revert to the brute *thingness* of the 'in-itself' (my p. 86) is a form of bad faith that Sartre portrays especially vividly, in instances ranging from the whimsical to the slightly ridiculous, and from the sinister to the deeply disturbing: from the grocer who does not dream because his customers expect a grocer to be 'wholly a grocer' [Sartre (1943), p. 59], to the café waiter with his mechanical, game-like behaviour serving the customers [ibid., pp. 59-60]; and from the soldier at attention who 'makes himself into a soldier-thing with a direct regard which does not see at all' [ibid., p. 59] to the anti-Semite who chooses to live in a state of hate and makes himself 'massive and impenetrable' because he wants to exist 'all at once and right away' [Sartre (1946), pp. 18-19]. Beyond my scope here, this is a theme I discuss and illustrate at greater length in Wilkinson (2012), pp. 88-90.

[331] Bohr (1928), p. 580. Quoted more fully on my p. 28.

[332] On the question how far quantum phenomena and effects may extend into the macroscopic world, see my n. 73.

[333] Nishida (1911), p. 3. Quoted more fully on my p. 69.

[334] See my n. 243.

[335] The *basho* of true nothingness is also Nishida's answer to the old puzzle of the one and the many, i.e., the question of why the world is so varied and mutable when there could just as well be a single and eternal Platonic Idea of a world: see my main text pp. 90-91.

[336] From Nishida's preface to '*Hataraku mono kara miru mono e*' ['From the Acting to the Seeing'], adapted from Krummel's quote in his introduction to Nishida (1926), p. 11. See my n. 225 for Nishida's original Japanese.

[337] Adapted from the Krummel/Nagatomo translation in Nishida (1926), p. 57; previously quoted on my p. 79.

# Notes

[338] Adapted from the Abe/Ives translation in Nishida (1911), p. 143; quoted more fully on my p. 82.

[339] Nishida (1911), p. 149; quoted earlier on my p. 80. See my n. 259 for the source of Nishida's phrase 'the consummation of philosophy'.

[340] My main text pp. 55-56.

[341] For the source of the Linji quote, see my n. 272. For an attempt to convey some of the nuances of the virtually untranslatable word '*byōjōtei*', my n. 271.

[342] As discussed on my pp. 71 and 80, Nishida uses 'true self' and related phrases interchangeably with 'personality', to some extent in the conventional sense of individual human characteristics but fundamentally as a synonym for 'the unity of consciousness'. For 'innate nature', see my pp. 81-82: as noted there, the bamboo and pine imagery is almost certainly an allusion to the poet Bashō.

[343] It is beyond my scope to explore whether this is what Nishida means by 'the eternal now [*eien no ima* or *eien no genzai*]', except to say that I suspect it is at least part of what he means.

[344] My main text p. 80. As noted there, the expression 'understood on the basis of reality' is Abe's, not Nishida's.

[345] Nishida (1911), p. 126; quoted more fully on my p. 80. Significantly, I think, Nishida originally suggested that Part III of the book, in which this passage occurs, could be regarded as 'an independent ethic [独立の倫理学]': ibid, Preface [1911] p. xxx; 同書、5頁。

[346] For the sources of these remarks and the context in which each was made (complementarity and the EPR thought-experiment in Bohr's case, the 'multitudinist' conception of the world in d'Espagnat's), see my pp. 35 and 95.

[347] See my n. 273 for the source of this quotation.

[348] Nishida (1911), p. 135; 西田 (1911)、205頁、「主客相没し物我相忘れ天地唯一実在の活動あるのみなるに至って、甫めて善行の極致に達するのである。物が我を動かしたのでもよし、我が物を動かしたのでもよい。雪舟が自然を描いたものでもよし、自然が雪舟を通して自己を描いたものでもよい。本来物と我と区別のあるのではない、客観世界は自己の反映といい得る様に自己は客観世界の反映である。我が見る世界を離れて我はない。」For Sesshū and the genre of *zenga* ('Zen painting'), see my n. 208.

[349] Adapted from the Krummel/Nagatomo translation in Nishida (1926), p.77; 西田 (1926)、111頁、「真の無の空間において描かれたる一点一画も生きた実在である。」

[350] Nishida (1911), pp. 130-31 and 3, in that order. Quoted more fully on my pp. 71 and 69.

www.ingramcontent.com/pod-product-compliance
Lightning Source LLC
Chambersburg PA
CBHW052145110526
44591CB00012B/1861